NITROGEN FLUXES IN INTENSIVE GRASSLAND SYSTEMS

Developments in Plant and Soil Sciences

1. J. Monteith and C. Webb, eds.,
Soil Water and Nitrogen in Mediterranean-type Environments. 1981. ISBN 90-247-2406-6
2. J.C. Brogan, ed.,
Nitrogen Losses and Surface Run-off from Landspreading of Manures. 1981. ISBN 90-247-2471-6
3. J.D. Bewley, ed.,
Nitrogen and Carbon Metabolism. 1981. ISBN 90-247-2472-4
4. R. Brouwer, I. Gašparíková, J. Kolek and B.C. Loughman, eds.,
Structure and Function of Plant Roots. 1981. ISBN 90-247-2510-0
5. Y.R. Dommergues and H.G. Diem, eds.,
Microbiology of Tropical Soils and Plant Productivity. 1982. ISBN 90-247-2624-7
6. G.P. Robertson, R. Herrera and T. Rosswall, eds.,
Nitrogen Cycling in Ecosystems of Latin America and the Caribbean. 1982. ISBN 90-247-2719-7
7. D. Atkinson et al., eds.,
Tree Root Systems and their Mycorrhizas. 1983. ISBN 90-247-2821-5
8. M.R. Sarić and B.C. Loughman, eds.,
Genetic Aspects of Plant Nutrition. 1983. ISBN 90-247-2822-3
9. J.R. Freney and J.R. Simpson, eds.,
Gaseous Loss of Nitrogen from Plant-Soil Systems. 1983. ISBN 90-247-2820-7
10. United Nations Economic Commission for Europe.
Efficient Use of Fertilizers in Agriculture. 1983. ISBN 90-247-2866-5
11. J. Tinsley and J.F. Darbyshire, eds.,
Biological Processes and Soil Fertility. 1984. ISBN 90-247-2902-5
12. A.D.L. Akkermans, D. Baker, K. Huss-Danell and J.D. Tjepkema, eds.,
Frankia Symbioses. 1984. ISBN 90-247-2967-X
13. W.S. Silver and E.C. Schröder, eds.,
Practical Application of Azolla for Rice Production. 1984. ISBN 90-247-3068-6
14. P.G.L. Vlek, ed.,
Micronutrients in Tropical Food Crop Production. 1985. ISBN 90-247-3085-6
15. T.P. Hignett, ed.,
Fertilizer Manual. 1985. ISBN 90-247-3122-4
16. D. Vaughan and R.E. Malcolm, eds.,
Soil Organic Matter and Biological Activity. 1985. ISBN 90-247-3154-2
17. D. Pasternak and A. San Pietro, eds.,
Biosalinity in Action: Bioproduction with Saline Water. 1985. ISBN 90-247-3159-3.
18. M. Lalonde, C. Camiré and J.O. Dawson, eds.,
Frankia and Actinorhizal Plants. 1985. ISBN 90-247-3214-X
19. H. Lambers, J.J. Neeteson and I. Stulen, eds.,
Fundamental, Ecological and Agricultural Aspects of Nitrogen Metabolism in Higher Plants. 1986. ISBN 90-247-3258-1
20. M.B. Jackson, ed.
New Root Formation in Plants and Cuttings. 1986. ISBN 90-247-3260-3
21. F.A. Skinner and P. Uomala, eds.,
Nitrogen Fixation with Non-Legumes. 1986. ISBN 90-247-3283-2
22. A. Alexander, ed.
Foliar Fertilization. 1986. ISBN 90-247-3288-3.
23. H.G. v.d. Meer, J.C. Ryden and G.C. Ennik, eds.,
Nitrogen Fluxes in Intensive Grassland Systems. 1986. ISBN 90-247-3309-x.
24. A.U. Mokwunye and P.L.G. Vlek, eds.,
Management of Nitrogen and Phosporus Fertilizers in Sub-Saharan Africa. 1986. ISBN 90-247-3312-x

Nitrogen Fluxes in Intensive Grassland Systems

Edited by

H.G. VAN DER MEER
Centre for Agrobiological Research (CABO)
Wageningen, The Netherlands

J.C. RYDEN†
The Animal and Grassland Research Institute
Hurley, Maidenhead,
Berkshire, U.K.

and

G.C. ENNIK
Centre for Agrobiological Research (CABO)
Wageningen, The Netherlands

1986 **MARTINUS NIJHOFF PUBLISHERS**
a member of the KLUWER ACADEMIC PUBLISHERS GROUP
DORDRECHT / BOSTON / LANCASTER

Distributors

for the United States and Canada: Kluwer Academic Publishers, 101 Philip Drive, Assinippi Park, Norwell, MA 02061, USA
for the UK and Ireland: Kluwer Academic Publishers, MTP Press Limited, Falcon House, Queen Square, Lancaster LA1 1RN, UK
for all other countries: Kluwer Academic Publishers Group, Distribution Center, P.O. Box 322, 3300 AH Dordrecht, The Netherlands

Library of Congress Cataloging in Publication Data

Nitrogen fluxes in intensive grassland systems.

(Developments in plant and soil sciences ;)
1. Grasses--Fertilizers. 2. Pastures--Fertilizers.
3. Nitrogen fertilizers. 4. Plants, Effect of nitrogen on. 5. Soils--Nitrogen content. 6. Nitrogen in animal nutrition. 7. Dairy farming. 8. Dairy farming--Netherlands. I. Meer, H. G. van der. II. Ryden, J. C. Ennik, G. C. IV. Series.
S667.G7N57 1986 633.2'894 86-2529

ISBN-13: 978-94-010-8454-3 e-ISBN-13: 978-94-009-4394-0
DOI: 10.1007/978-94-009-4394-0

Book information

Legend to cover photograph: Equipment for measuring the volatilisation of ammonia from grazed plots at Hurley (micrometeorological mass balance method; page 60 of this volume) (photograph: The Animal and Grassland Research Institute, Hurley, UK).

John Christopher Ryden
20 October 1947–3 May 1986

This book is dedicated to the memory of Dr. John Christopher Ryden.

When the corrected proofs of this book were about to be returned to the publishers, we heard of the sudden death of our co-editor, John Ryden. His many colleagues and friends were deeply shocked by this tragic news.

John Ryden was one of the world's leading research workers concerned with nitrogen flows in grassland systems. He worked on nitrogen for a number of years in the United States of America and in the United Kingdom. He developed methods for measuring nitrogen flows in the field and he quantified the various losses which take place within a grazed pasture. The two papers included in this volume of which he was author and co-author provide evidence of his great expertise in this difficult area of agricultural science. During his short career he was remarkably productive, both in terms of research carried out and publications written. He advocated the application of whole-system measurements of nitrogen flows in grassland under different conditions, and was a valued adviser for the implementation of such measurements, e.g. in a large scale experiment at the Ir. A.P. Minderhoudhoeve in the Netherlands. During Summer 1985 he visited a number of grassland research centres in China.

We got to know John well during the EEC-Workshop "Nitrogen fluxes in intensive grassland systems" at Wageningen in 1983 and particularly in the selection and editing of papers for this volume. Our admiration for his was enhanced during a visit that he made to the Netherlands in October 1985. The purpose of this visit was to gain information about Dutch research on the utilization of animal slurries and the volatilization of ammonia. In this context he spoke with many research workers and delivered several talks on his own work. It was a very inspiring and pleasant experience to accompany John during the main part of this visit.

Apart from being an outstanding scientist John Ryden was a very amiable man, who at first meeting gave the impression of being rather shy, but soon showed himself to be a sincere and easy going person with whom it was good to work, and also to spend leisure time. Hiking was his main leisure activity and, despite the overcrowded programme during his visit to the Netherlands last year, we enjoyed an unforgettable, relaxing Sunday with a bike-ride through woods and heathlands to the Kröller-Muller Museum.

The Animal and Grassland Research Institute at Hurley has lost an outstanding reserach scientist and the scientific world an important contributor.

Hugo G. van der Meer
Gerrit C. Ennik

Wageningen,
May 1986

Contents

The relationship between inputs and outputs of nitrogen in intensive grassland systems

HG VAN DER MEER and MG VAN UUM-VAN LOHUYZEN
Centre for Agrobiological Research (CABO), P.O. Box 14, 6700 AA Wageningen, The Netherlands

Key words: grassland, nitrogen balance, nitrogen losses, soil nitrogen, fertilizer nitrogen, nitrogen recovery, nitrogen efficiency

Abstract. On intensively managed grassland farms, large amounts of nitrogen (N) are introduced in fertilizer and concentrates. These inputs contribute significantly to the current productivity of grassland and livestock. However, the large difference between inputs of N in fertilizer and purchased feeds and outputs in animal products, e.g. up to 450 kg N ha^{-1} yr^{-1} on a group of well managed, intensive dairy farms in The Netherlands, indicates a potentially large emission of N to the environment. It is emphasized that studies on the fate of N in grassland systems should consider the whole production system.

The response of herbage yield to applied N is an important aspect of the efficiency of N utilization on grassland farms. The large differences in this response, observed in experiments, arise mainly from differences in N supply from sources other than fertilizer and in apparent recovery of ferilizer N in the herbage. Results from N manuring experiments on grassland in The Netherlands indicate that the recovery of fertilizer N in the herbage has increased since 1970.

It is shown that the supply of N from sources other than fertilizer significantly affects the optimum rate of fertilizer N. Prediction of the supply of N from soil and animal excreta may lead to a substantial improvement in the precision of the fertilizer recommendations and, hence, to improved efficiency of N use on intensive grassland farms.

As early as the beginning of this century, field trials by the Dutch Advisory Service had demonstrated that grass yields could be increased by applying nitrogen (N) [18]. It was not until the thirties, however, when large scale production resulted in low prices of fertilizer N, that farmers started to apply N to grassland. After World War II, favourable price relationships stimulated the use of N [9, 18] and the average rate of application of fertilizer N on the 1.2 million ha of grassland in The Netherlands, increased steadily from 50 kg ha^{-1} in 1950 to 265 kg ha^{-1} in 1980 [10]. This is still lower than the recommended rates for grassland, which are about 400 kg N ha^{-1} yr^{-1} for grassland on sand, clay and wet peat and 250 kg N ha^{-1} yr^{-1} on well drained peat soils [8, 39]. However, the figures on actual use refer only to the fertilizer N applied, whereas the recommendations also include an estimate of the amount of available N in slurry [39, 40]. Hence, the difference between actual and recommended rates of N application is smaller than the figures suggest and many farmers apply the recommended rate or even more [37].

H G van der Meer, J C Ryden and G C Ennik (eds) Nitrogen Fluxes in Intensive Grassland systems.
 ISBN 978-94-010-8454-3

The use of fertilizer N on grassland has also increased in other countries of the European Community. For instance, in the United Kingdom rates increased from 15 to 130 kg N ha^{-1} yr^{-1} between 1950 and 1980 [42]. Although in other countries the amounts of N applied per ha are on average lower than in The Netherlands [10], up to 450 kg N ha^{-1} yr^{-1} are used to sustain production on the most intensive dairy farms [13, 42].

The increased input of N has made an important contribution to the improvement of grassland productivity. However, in 1973 the price of fertilizer N started to rise, mainly due to a rapid increase in the cost of fossil energy [34]. Furthermore, during the last decade there has been growing concern about the impact of high N inputs on the natural environment. This has related mainly to leaching of nitrate to groundwater and surface water [5, 14, 46, 53] and to volatilization of ammonia, which after deposition may lead to soil acidification and undesired enrichment of nature reserves [7]. Both economic and environmental concerns demand a better understanding of the fate of N in intensive grassland systems in order to increase the efficiency with which N is used, without adverse effects on productivity.

The potential for N losses from a farming system can be estimated from the N budget for that system [16]. In grassland, this is achieved in the first place at farm level by balancing inputs of N in fertilizer, biological fixation, purchased feeds and precipitation against outputs in animal products. However, as fertilizer N is by far the most important N input on intensive grassland farms in Western Europe, it is also necessary to consider the efficiency of use of fertilizer N in herbage production. The purpose of this paper is to assess the relationships between inputs and outputs of N in intensive grassland systems at both the whole farm and crop levels.

Nitrogen balances of intensive grassland farms

A study of the N balances of 14 intensive dairy farms in The Netherlands showed that only 16% of the total N input in fertilizer, purchased feeds and precipitation (equivalent to 533 kg N ha^{-1} yr^{-1}) was removed in animal products [27]. In the year of this study, these farms applied on average 383 kg fertilizer N per ha grassland plus fodder crops and bought 1575 kg dry matter (DM) of concentrates plus 675 kg DM of roughage per milking cow. Hence, the average rate of N application was broadly similar to the recommended rate and the consumption of purchased feeds was similar to the national average [25]. Animal production on these farms was 13 500 kg milk plus approximately 470 kg liveweight gain per ha per year.

The fraction of the N input which was accounted for in animal products from the farms studied, ranged from 12.5 to 19%. Unfortunately, the number of farms included in the study was too small and the factors involved too numerous to analyse the causes of this variation. However, based on an analysis of the effect of fertilizer N on the N cycle of grassland farms, some causes of the inefficient use of N and possibilities for improvement can be identified [27, 28]. Fig. 1 illustrates the flow of N on these intensive dairy farms. The annual N yield in the accumulated herbage is estimated at 400 kg ha^{-1}, of which 20% is left on the land by grazing and harvesting losses. The diagram shows clearly the inefficient utilization of N by the animals: only 20% of the consumed N is retained in milk and liveweight.

The data in Fig. 1 indicate that the difference between total N input and N output in animal products is 440 kg ha^{-1} yr^{-1}. The fate of this large amount of N is unknown. In grassland systems with an equilibrium or 'steady-state' level of organic N in the soil, where addition of organic N equals mineralization, the amount of N not recovered in animal products will be lost by a combination of volatilization of ammonia, denitrification, leaching and runoff. As permanent grassland prevails on the farms studied, the content of organic N in the soil is expected to be at or close to an equilibrium [48]. However, aspects of management associated with intensification of production may be affecting the amount and quality of added organic matter and/or the mineralization rate and, in turn, the content of organic N in the soil. These include increased use of N and purchased feeds, improved

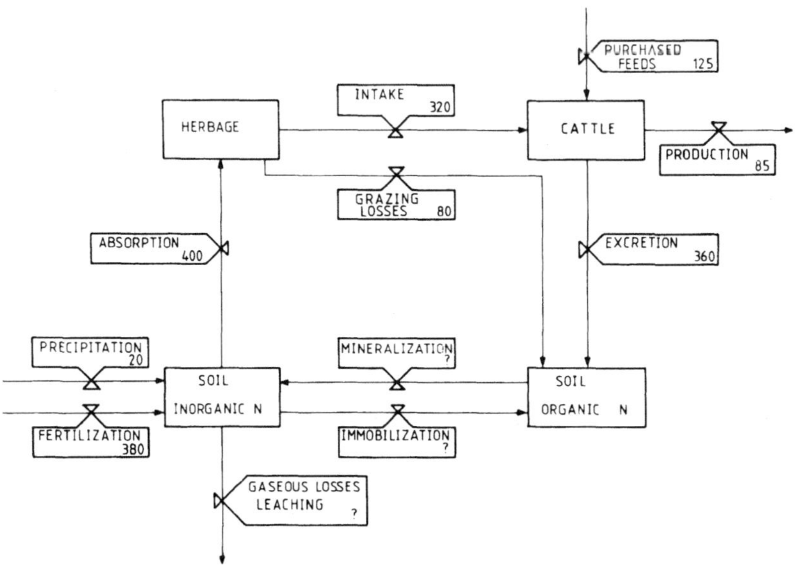

Figure 1. Flow diagram for N in intensive dairy farms in The Netherlands, presented according to Penning de Vries [35]. Numerical values are kg N ha^{-1} yr^{-1}.

herbage utilization, increased frequency of pasture renovation (with soil cultivation), decreased use of bedding in animal housing, improved drainage and a lowered watertable in many grassland regions. No information is available on the combined effect of such developments on the N status of permanent grassland soils under Western European conditions. Hence, the actual N losses on the farms studied may be greater or smaller than 440 kg ha^{-1} yr^{-1}, depending whether the soils are accumulating or losing organic N.

Fig. 1 shows that the N input in purchased feeds is higher than the N output in animal products. This is common on Dutch dairy farms at the present average concentrate consumption of about 2000 kg DM per milking cow per year with a mean N content of 28 g per kg DM [25, 47]. Hence, on farms with this level of concentrate feeding the difference between total N input and N output in animal products (the potential loss of N) is slightly greater than the N input in fertilizer, precipitation and biological fixation.

To date, most information on N losses from grassland has been obtained using cut swards [e.g. in The Netherlands: 1, 23, 24, 36, 55]. This information has led to the opinion that up to the recommended rate of N application (about 400 kg ha^{-1} yr^{-1}), the risk of N losses to the environment, especially nitrate leaching, is insignificant [9, 10, 37]. Under normal conditions this certainly holds for cut swards, but there is increasing evidence that the situation is quite different if the whole production system, including the animals, is considered [2, 3, 4, 20, 27, 33, 42, 48]. Several companion papers in this volume [3, 17, 43, 50] deal with effects of the ruminants on the fate of N in intensive grassland systems.

Nitrogen input/output ratios in grass production

Fertilizer N has been called 'the most effective management input for manipulating grassland yield within the limitations imposed by the environment' [32]. As N enhances grass growth, the grass can be grazed or cut at a younger stage of growth and consequently will have a higher feeding value [19, 30, 52]. However, the effect of applied N on the content of metabolizable or net energy is rather small compared to the effect on digestible crude protein [52]. This leads to an increasing surplus of protein (or relative shortage of metabolizable energy) in the herbage [22, 52], which is one of the causes of the low recovery of N in animal products on intensively fertilized pastures [27]. Hence, from the viewpoint of N efficiency, the effect of applied N on the yield (and content) of metabolizable energy has to be considered rather than the effect on herbage N yield. As the effect of N on the content of metabolizable energy in the grass is small (under good management), the effect on the yield of metabolizable energy is broadly similar to the effect on DM yield.

Many authors have pointed to the large differences in the response of herbage yield to fertilizer N [e.g. 10, 30, 31, 48, 52]. According to De Wit [15, 54], the response of DM yield to fertilizer N results from the combined effects of applied N on N uptake by the crop (or N yield in the harvested parts of the crop) and of N uptake on DM yield. These interrelationships are conveniently summarized in the 'three quadrants diagram', which can be used to analyse the effect of fertilizer N on yield, and interactions between fertilizer N and other production and management factors including soil moisture, disease control and cutting frequency [12, 27, 44]. Use of this diagram shows [27] that differences in the response of herbage yield to fertilizer N result from differences in:

- the supply of N from sources other than fertilizer, usually measured as the N yield in the herbage from swards without fertilizer N;
- the apparent recovery of fertilizer N in the herbage ('apparent recovery' is calculated as the difference in N yield between a fertilized sward and an untreated sward, expressed as percentage of the N applied);
- the DM production per kg of N harvested in the herbage.

These aspects of the response to fertilizer N are discussed below.

The nitrogen yield in the herbage from swards without fertilizer nitrogen

From an analysis of the results of N manuring experiments on grassland in The Netherlands (H.G. van der Meer and M.G. van Uum-van Lohuyzen, unpublished), it was apparent that the N yield of swards without fertilizer N was in most cases between 100 and 250 kg ha^{-1} yr^{-1}. The results from each experiment in different years were considered separately, giving a total of about 400 experiment-years. However, high and low values occurred rather frequently; the extremes were 40 kg N ha^{-1} in the fourth production year of a *Lolium perenne* ley on heavy clay [36] and 528 kg N ha^{-1} yr^{-1} on permanent grassland on a peat soil after the watertable was lowered [6].

The N yields on unfertilized swards in Dutch experiments are on average considerably higher than those reported in the UK [31, 41]. The main cause of this difference probably arises from the fact that many N experiments in The Netherlands have been carried out on intensively managed permanent grassland, whereas experiments in the UK have often used young swards on previously arable land.

The large range in N supply from the soil, combined with the recommended high rates of N application for grassland in The

Netherlands, may lead to a total N supply which exceeds the requirements for crop growth and results in uneconomic use of fertilizer N and increased potential for N loss. At present, fertilizer recommendations for grassland in The Netherlands take account only of the high mineralization of N occurring in well drained peat soils. However, available information suggests that a prediction of the soil N supply may lead to a substantial improvement in the precision of fertilizer recommendations.

The main sources of available soil N on grassland in the growing season are: (i) inorganic N in the soil at the start of the growing season, (ii) N mineralized from soil organic matter, and (iii) in grazed swards a portion of the N returned in animal excreta and herbage residues (Fig. 1). The soil under intensively managed grassland generally contains inorganic N in early spring [27]. On 57 grassland farms sampled in February and March between 1978 and 1980, Leusink [26] found on average 63 kg mineral N per ha (range: 13–348) in the soil layer of 0–30 cm. In N manuring experiments, conducted by CABO between 1979 and 1981 at different sites on well-drained sand and clay soils, a close relationship was observed between the amount of mineral N (N_{min}) in the root zone of the soil profile in early spring and the N yield in the herbage (N_y) cut from nil fertilizer N plots in the following season (Fig. 2):

$$N_y = 0.66\,N_{min} + 114 \qquad r = 0.841 \qquad n = 18 \qquad (1)$$

The data relate to soils collected from both intensively and extensively managed swards. In these experiments, the average amount of N_{min} in early spring was 104 kg ha^{-1} (range: 11–263). At the end of the growing season, N_{min} in the soil of the unfertilized plots was on average 27 kg ha^{-1} (range: 7–74).

The prediction of N supply from the soil in these experiments was improved by including an estimate of mineralizable N (N_{minz}), which was determined by incubating soil samples for 2 weeks at 30 °C (Fig. 3):

$$N_y = 0.67\,(N_{min} + N_{minz}) + 21 \qquad r = 0.901 \qquad n = 18 \qquad (2)$$

It is interesting to note that in both figures the slope of the line is similar (0.66 and 0.67 in the equations (1) and (2), respectively). The slope indicates the recovery in herbage of the different sources of soil N; the value of 0.67 is comparable to the apparent recovery of fertilizer N which averaged about 70% in these experiments. Furthermore, it is worth noting that the intercept in Fig. 2 compares well with 67% of the average amount of N_{minz}, which was 137 kg ha^{-1} in these experiments (range: 79–203), whereas the line in Fig. 3 passes close to the origin.

Prediction of N supply from the soil by determination of

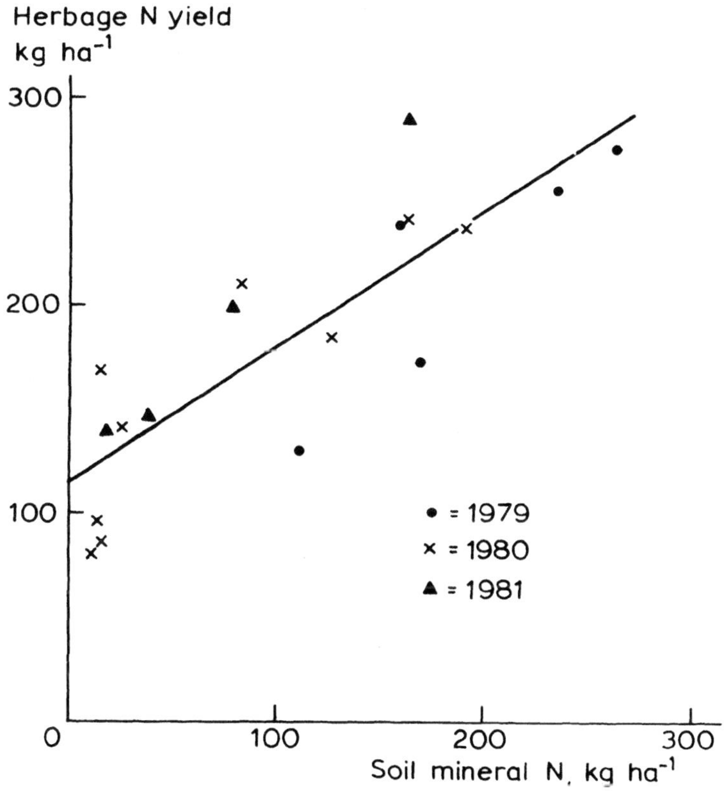

Figure 2. Effect of the amount of mineral N in the root zone of soil profiles in early spring on N yield in herbage cut from nil fertilizer N plots in the following season. At each site, fifteen cores to depths of 60–100 cm were collected, the maximum sampling depth being determined by the rooting depth.

$N_{min} + N_{minz}$, proved to be unsuccessful on peat soils with a high and variable watertable. In these soils, mineralization of N depends on the level of the watertable during the growing season, which is strongly affected by the weather.

Despite the success of the approach outlined above, estimation of N_{minz} by the incubation method is too time-consuming for practical use. Alternative, more rapid methods need to be developed [49, 51]. Furthermore, attempts should be made to develop simulation models of the mineralization of organic N in intensive grassland systems.

Recovery of fertilizer nitrogen by grass

Only a part of the applied fertilizer N is harvested in the herbage, and this is usually expressed as 'apparent recovery of N'. Woldendorp *et al.* [55] stated at the First General Meeting of the European Grassland

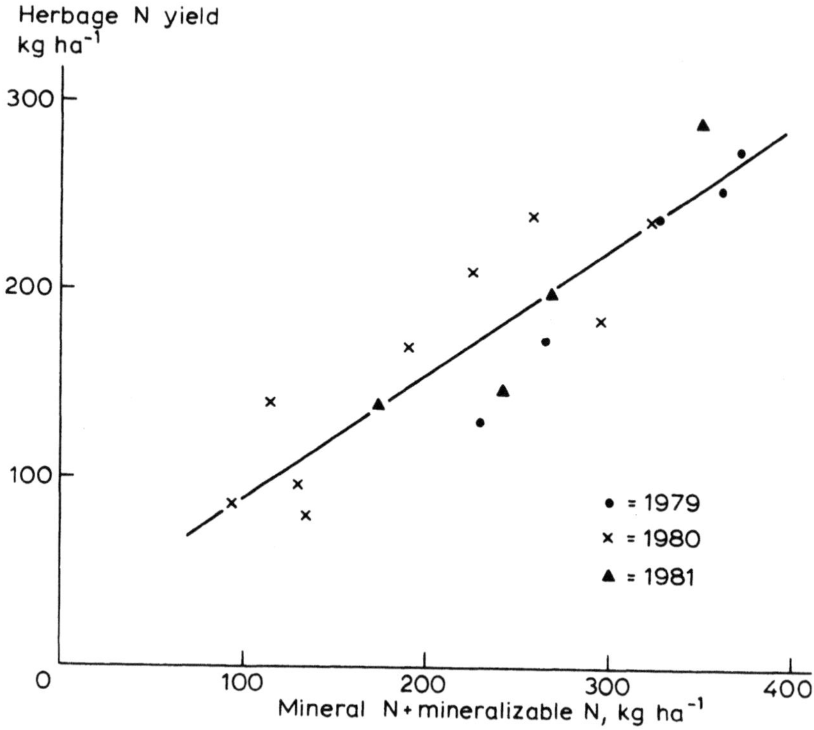

Figure 3. Effect of mineral N plus mineralizable N in the root zone of soil profiles in early spring on N yield in herbage cut from nil fertilizer N plots in the following season. The soils used were the same as those used to obtain Fig. 2.

Federation: 'It is a general experience that on permanent grassland soils in The Netherlands, the recovery of fertilizer nitrogen in the herbage seldom exceeds 60 per cent'. This was confirmed in a large series of N manuring experiments on permanent grassland in the period 1964–1973 [27, 45]. However, in more recent experiments we have found considerably higher values. This prompted us to examine whether the apparent recovery of fertilizer N applied to grassland had improved during the past 15 to 20 years. For this purpose N recoveries from applications of 300 kg N ha^{-1} yr^{-1} were compared. This rate of application was chosen because the corresponding recovery of fertilizer N is generally close to its maximum and less variable than at lower rates [11, 27]. Results were omitted in the case of experiments in which the highest rate of applied N was less than 300 kg ha^{-1} yr^{-1}. If no 300 but a 400 kg N treatment had been included, the recovery of N from 400 kg ha^{-1} was used. For experiments in which levels of N supply were combined with other experimental treatments, the average N recovery from 300 kg N over each of the treatments was calculated and used in the

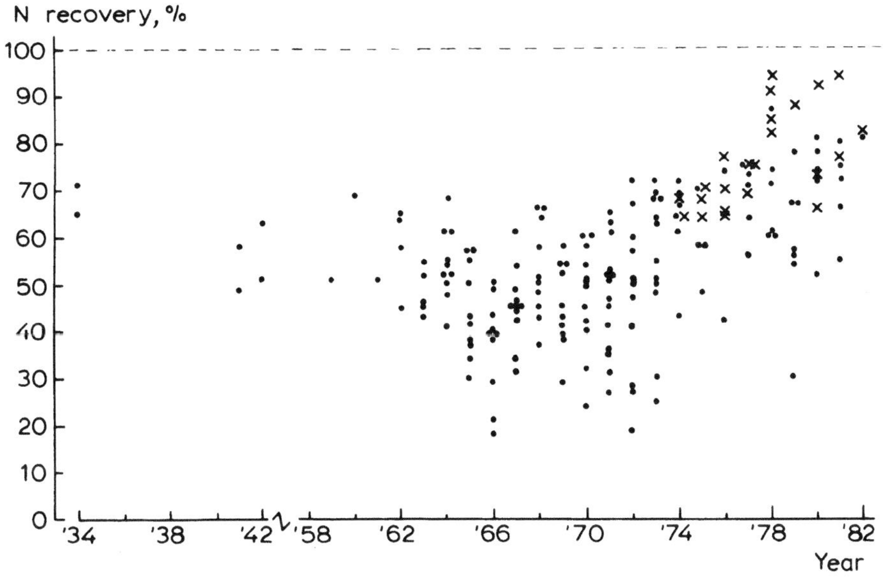

Figure 4. Trend of apparent recoveries of fertilizer N applied to grassland in The Netherlands. Dots and crosses indicate apparent recoveries of N in herbage at application rates of 300 or 400 kg N ha^{-1} yr^{-1}, respectively.

analysis of data. Recoveries of N from the three experimental fields on each soil type in the PAW 970 series [45] have also been averaged. Results from each experiment in different years were considered independently. Altogether, 192 suitable experiment-years were available; the resulting N recoveries are presented in Fig. 4.

The data show that the recovery in herbage of fertilizer N applied to grassland experiments has improved since 1970. After that year, N recovery was over 70% in an increasing number of experiments; before 1970 this occurred only once. The causes of this improvement are not yet fully understood.

Analysis of the results of the PAW 970 experiments showed that very low recoveries of fertilizer N (under 40%) occurred at wet sites, especially in years with a wet growing season [27]. Minderhoud [29] reported an improvement of the N recovery after lowering the water-table on a heavy clay soil. However, the opposite was found on peat [6], due to the strongly increased mineralization of soil N after drainage. In the last decade the number of N experiments on wet grassland sites decreased, in part as a consequence of the improvement of drainage in many grassland regions, but also as a result of changing research topics. Another important factor is the presence of some white clover in part of the early N manuring experiments, whereas clover-free swards were mostly used in the last decade. Low apparent recoveries of N have been found on grass-clover swards owing to the decline of legume content and N fixation at increasing rates of N fertilizer [27]. These two factors, *viz.* improved drainage and less clover in the swards may explain the small number of low N recoveries in grassland experiments after 1973.

However, it seems more difficult to explain why very high recoveries of fertilizer N were not found before the 1970's, as studies before this date also included pure grass swards on well-drained soils. A factor which may be involved, is the probable decrease of the C/N ratio of resident organic matter in the soil as a consequence of intensification of pasture production and utilization [2]. This may lead to decreased immobilization of applied N and therefore to a higher recovery in the herbage. This point needs further research. Additionally, in the course of time, there may also have been a small positive effect arising from the general improvement of grassland husbandry, e.g. through increased supply of other plant nutrients and a greater content of grasses in the sward capable of efficient utilization of applied N. In this context, it is worth noting that Jenkinson [21] reported a strong increase of the apparent recovery of fertilizer N in wheat in the Rothamsted long-term cereal experiments which he attributed to new, high-yielding varieties, to better control of pests and diseases and to earlier sowing in autumn. However, on grass, these factors mainly improve the production capacity of the sward which finds expression in a higher DM yield at high N uptakes and only a small increase of the N recovery, mainly at high rates of fertilizer N [12] (G.C. Ennik, private communication).

Relationship between nitrogen uptake and herbage yield

In general, N uptake is measured as the N uptake or N yield in the harvested parts of the crop. In experiments with a similar cutting frequency, the relationship between N yield of $<300\,\mathrm{kg\,N\,ha^{-1}\,yr^{-1}}$ and herbage DM yield is identical (T. van Steenbergen, unpublished) [28]. At higher N yields, the DM response tends to level off and this occurs most rapidly under unfavourable growing conditions, e.g. moisture deficit. The curvilinear relationship between N yield and DM yield demonstrates the decreasing DM response and the increasing N content in the herbage at increasing N supply. Both points affect the N efficiency of the whole production system [27].

The relationship between N yield and DM yield is strongly affected by cutting frequency: at the same N yield, DM yield is higher with fewer cuttings per year [37, 44]. This does not necessarily mean, however, that at a lower cutting frequency N is used more efficiently.

Cutting frequency also affects feeding value and in this sense N efficiency should be considered in relation to the yield of metabolizable energy and its content in the DM rather than to the DM yield *per se*.

Variation in the response of herbage yield to fertilizer nitrogen

As discussed above, the response of herbage yield to fertilizer N depends on the uptake of 'soil N', the apparent recovery of fertilizer N in the herbage and the DM production per kg of N taken up by the crop. Growing conditions and pasture management generally affect one or more of these three aspects of N utilization. This can be analysed by means of the 'three quadrants diagram', which relates supply of fertilizer N, N uptake in the herbage and DM yield (see Fig. 5). In the 'three quadrants diagram', quadrant IV shows the relationship between supply of fertilizer N and N uptake in the herbage (the intercept indicates the N uptake at nil fertilizer N and the slope of the line the apparent recovery of fertilizer N in the herbage); quadrant I shows the relationship between N uptake in the herbage and DM yield and quadrant II the relationship between supply of fertilizer N and DM yield.

In the discussion below, the 'three quadrants diagram' is used to analyse the separate effects of each of the three aspects of N utilization namely, soil N supply, N recovery and the relationship between N uptake and DM yield, on the response of DM yield to applied fertilizer N (Figs. 5, 6 and 7, respectively). In each figure one aspect is varied and the other two are kept constant. The relationships in these figures are hypothetical, but are typical of those observed in N manuring experiments on grassland.

Fig. 5 shows the effect of two levels of 'soil N', resulting in the uptake in herbage of 100 and 250 kg N ha^{-1} yr^{-1} (intercepts in quadrant IV), on the response of DM yield to applied fertilizer N (solid and dashed lines, respectively). Both levels of 'soil N' are combined with identical relationships between supply of fertilizer N and N uptake (parallel lines for N uptakes >250 kg ha^{-1} yr^{-1}; quadrant IV) and between N uptake and DM yield (quadrant I). By this, the difference in uptake of 'soil N' is transformed into a difference in DM response to applied fertilizer N (quadrant II). The relationship between N uptake and DM yield in quadrant I of this figure is typical of many experiments in which the herbage is cut six times per year.

Fig. 6 shows the effect of two different apparent recoveries of fertilizer N, *viz*. 50 and 80% (solid and dashed lines respectively), such as may arise from differences in soil drainage, on the response of DM yield to applied fertilizer N. For this, both recovery levels are combined with the same uptake of 'soil N' in the herbage (150 kg N ha^{-1} yr^{-1}) and

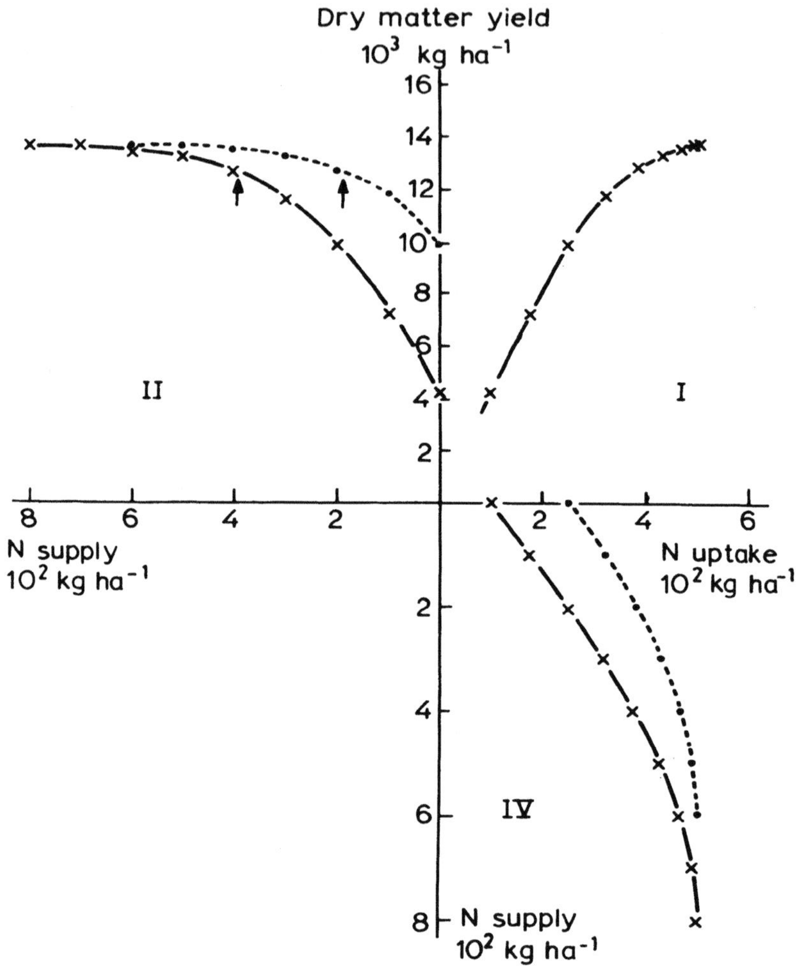

Figure 5. Effect of different levels of 'soil N' (quadrant IV) on the response of herbage DM yield to fertilizer N supply (quadrant II). Solid and dashed lines correspond to N uptakes from the soil (nil fertilizer N supply) of 100 and 250 kg ha^{-1} yr^{-1}, respectively.

an identical relationship between N uptake and DM yield. The recoveries of N correspond to the amount of fertilizer N needed to give a total N yield in the herbage of 450 kg ha^{-1} yr^{-1}. The relationship between N uptake and DM yield is the same as that in Fig. 5.

Fig. 7 shows the effect of a different relationship between N uptake in the herbage and DM yield on the response of DM yield to fertilizer N supply. A difference similar to that assumed in quadrant I was found in an irrigation experiment, but similar differences may arise from the use of different cultivars (G.C. Ennik, private communication) or

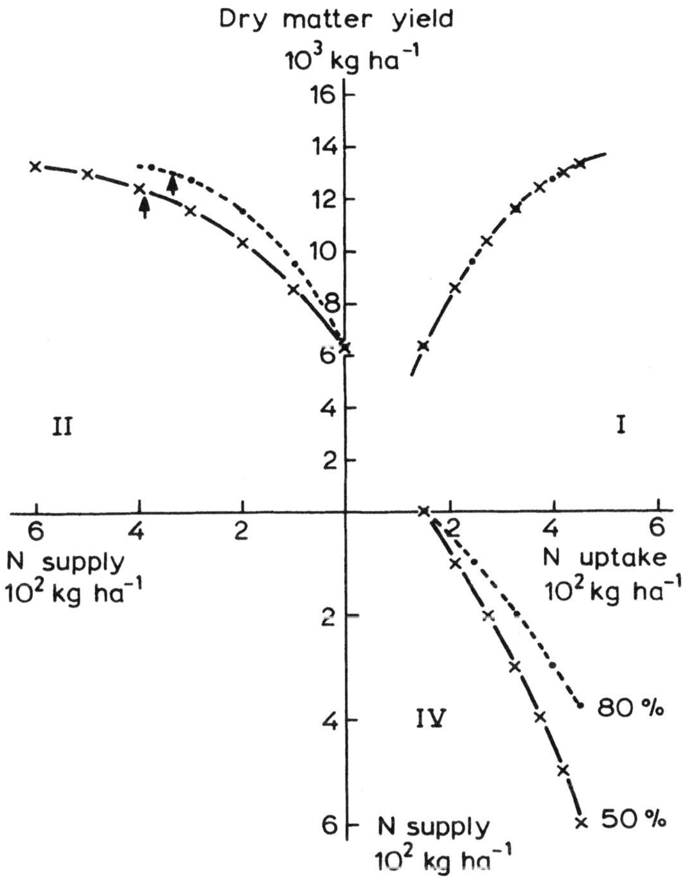

Figure 6. Effect of different apparent recoveries of fertilizer N (quadrant IV) on the response of herbage DM yield to fertilizer N supply (quadrant II). Apparent recoveries of 50 and 80% (solid and dashed lines, respectively) correspond to the amount of fertilizer N needed to give a N yield in the herbage of 450 kg ha^{-1} yr^{-1}.

imposed disease control [12]. Each relationship in quadrant I is combined with an uptake of 'soil N' of 150 kg ha^{-1} yr^{-1} and identical recoveries of applied fertilizer N.

Arrows in quadrant II of Figs. 5, 6 and 7 indicate the optimum level of fertilizer N supply assuming an economic marginal N effect of 7.5 kg DM per kg N [38]. In Fig. 5, the arrows show that differences in uptake of 'soil N' strongly affect the optimum rate of fertilizer N, illustrating the importance of a good prediction of available soil N. In Fig. 6, the arrows show a surprisingly small effect of an improved recovery of fertilizer N on the optimum level of fertilizer N. However, as an improved recovery of fertilizer N results in a steeper response of

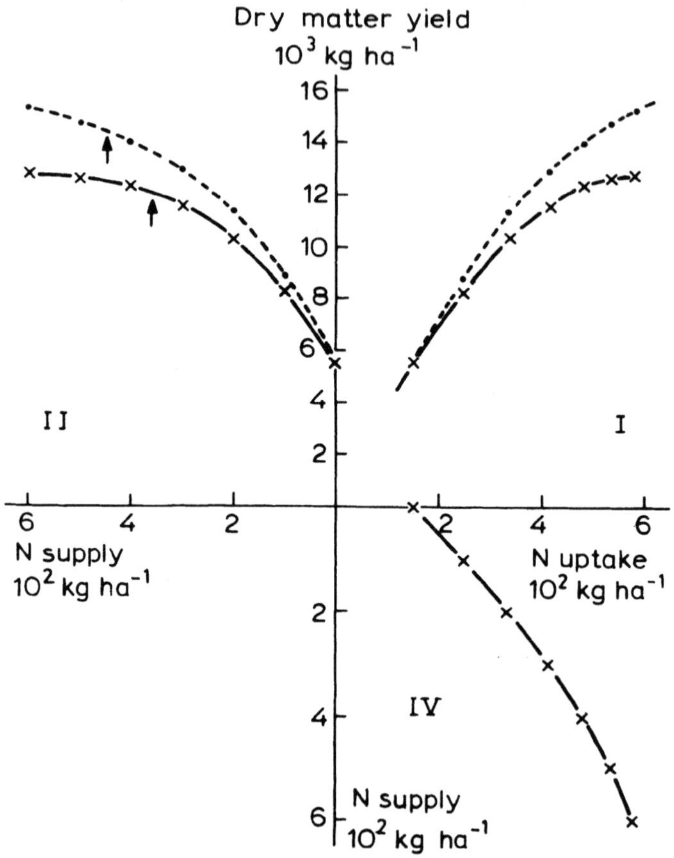

Figure 7. The response of herbage DM yield to fertilizer N supply (quadrant II) as affected by different relationships between N uptake and DM yield (quadrant I). Solid and dashed lines correspond to natural rainfall and irrigation treatments, respectively.

DM yield to applied fertilizer N, the optimum rate of fertilizer N is at a slightly higher DM yield. An improvement of DM yields at high levels of N uptake, improves the response of DM yield to fertilizer N and results in a higher optimum rate of fertilizer N, as indicated by the arrows in Fig. 7.

Efficiency of use of fertilizer nitrogen

Figs. 5, 6 and 7 can also be considered in relation to the efficiency of N use at the farm level (whole production system) and be used to identify methods to improve this efficiency. At an equilibrium level of organic N in the soil, the ratio between DM yield at optimum N supply (indicated by the arrows in the figures) and the amount of fertilizer N required to attain this yield, is an important measure of N efficiency at the farm level. Fig. 5 clearly indicates that this ratio is greatly improved by management practices which increase the uptake of N from sources other than fertilizer (N from soil, animal excreta and unharvested herbage and roots). In contrast, measures which improve the apparent recovery of fertilizer N in the herbage produce only a small improvement of the efficiency of N use (Fig. 6), whereas measures which improve the production capacity of the sward (e.g. irrigation, improved grass varieties) do not necessarily improve the N efficiency as they also result in a higher optimum rate of N application (Fig. 7).

Conclusions

The data reviewed in this paper indicate that on intensively managed grassland farms the difference between inputs of N in fertilizer and purchased feeds and outputs in animal products may amount to 450 kg N $ha^{-1} yr^{-1}$. As a consequence, these farms have a potentially large emission of N to the environment. The difference between inputs and outputs of N arises mainly from the effect of the ruminant on the utilization of N: only 20% of the consumed N is retained in milk and liveweight and, with present management systems, N excreted in urine and dung is poorly utilized in herbage production.

As fertilizer N is by far the most important input of N, the response of herbage yield to applied N is an important aspect of N efficiency on intensive grassland farms. This response varies widely, mainly as a result of differences in N supply from sources other than fertilizer ('soil N' and N from animal excreta) and in apparent recovery of fertilizer N in the herbage. Both points affect the optimum rate of application of fertilizer N and the differences observed suggest that greater precision is required in N fertilizer recommendations for grassland. This could be based on estimates of:

– the N uptake by the crop at optimum N supply,
– the N supply from the soil and animal excreta, and
– the apparent recovery of applied fertilizer N.

References

1. Alberda Th (1972) Nitrogen fertilization of grassland and the quality of surface water. Stikstof (English edn) 15, 45–51
2. Ball P R (1982) Nitrogen balances in intensively managed pasture systems. In Gandar P W, ed. Nitrogen balances in New Zealand ecosystems, pp 47–66. Palmerston North, New Zealand: DSIR
3. Ball P R (1983) Effects of grazing ruminants on the nitrogen economy of pastures. Report EEC–Workshop on Nitrogen fluxes in intensive grassland systems, pp 83–100. Wageningen, The Netherlands: CABO

4. Ball P R and Keeney D R (1981) Nitrogen losses from urine-affected areas of a New Zealand pasture, under contrasting seasonal conditions. In Smith J A and Hays V W, eds. Proceedings of the XIV International Grassland Congress (Lexington), pp 342–344. Boulder, Colorado: Westview Press
5. Bleuten W and Cerutti M (1984) De huidige en toekomstige nitraat- en sulfaatbelasting van grond- en drinkwater van de Nederlandse pleistocene zandgronden. H_2O 17, 208–212+ 218
6. Boxem Tj and Leusink A W F (1978) Ontwatering van veengrasland, verslag van een vergelijkend onderzoek onder bedrijfsomstandigheden te Zegveld van 1970 t/m 1975. Proefstation voor de Rundveehouderij, Publikatie No. 11. Lelystad, The Netherlands: PR
7. Breemen N van, Burrough P A, Velthorst E J, Dobben H F van, Wit T de, Ridder T B and Reijnders H F R (1982) Soil acidification from atmospheric ammonium sulphate in forest canopy throughfall. Nature 299, 548–550
8. Burg P F J van (1977) Nitrogen fertilization of grassland. In Gilsenan B, ed. Animal production from temperate grassland. Proceedings of an international Meeting (Dublin), pp 104–108
9. Burg P F J van, Hart M L 't and Thomas H (1980) Nitrogen and grassland – past and present situation in the Netherlands. In Prins W H and Arnold G H, eds. The role of nitrogen in intensive grassland production. Proceedings of an international Symposium of the European Grassland Federation (Wageningen), pp 15–33. Wageningen, The Netherlands: PUDOC
10. Burg P F J van, Prins W H, Boer D J den and Sluiman W J (1981) Nitrogen and intensification of livestock farming in EEC countries. Proceedings No. 199. London, UK: The Fertiliser Society
11. Dowdell R J, Morrison J and Hood A E M (1980) The fate of fertilizer nitrogen applied to grassland: uptake by plants, immobilisation into soil organic matter and losses by leaching and denitrification. In Prins W H and Arnold G H, eds. The role of nitrogen in intensive grassland production. Proceedings of an international Symposium of the European Grassland Federation (Wageningen), pp 129–136. Wageningen, The Netherlands: PUDOC
12. Ennik G C (1972) Dry matter yield response of pasture grass to application of Nemafos (thionazin). Netherlands Journal of Agricultural Science 20, 81–96
13. European Grassland Federation and British Grassland Society (1982) Efficient grassland systems in practice. In Corrall A J, ed. Efficient grassland farming. Proceedings of the 9th General Meeting of the European Grassland Federation (Reading), pp 93–108. Occasional Symposium No. 14 of the British Grassland Society. Hurley, UK: The British Grassland Society
14. Foster S S D, Cripps A C and Smith-Carington A (1982) Nitrate leaching to groundwater. Philosophical Transactions of the Royal Society (London) 296, 477–489
15. Frankena H J and Wit C T de (1958) Stikstofbemesting, stikstofopname en grasgroei in het voorjaar. Landbouwkundig Tijdschrift 70, 465–472
16. Frink C R (1969) Water pollution potential estimated from farm nutrient budgets. Agronomy Journal 61, 550–553
17. Garwood E A and Ryden J C (1986) Nitrate loss through leaching and surface runoff from grassland: effects of water supply, soil type and management (this volume)
18. Groot H de (1965) Nitrogen fertilization in the Netherlands. In Burg P F J van and Arnold G H, eds. Nitrogen and grassland. Proceedings of the 1st General Meeting of the European Grassland Federation (Wageningen), pp 27–38. Wageningen, The Netherlands: PUDOC
19. Hartmans J (1975) The effect of intensification on crop composition and on health and production of livestock. Stikstof (English edn) 18, 12–20
20. Henzell E F (1970) Problems in comparing the nitrogen economics of legume-based and nitrogen-fertilized pasture systems. In Norman M J T, ed. Proceedings of the XI International Grassland Congress (Surfers Paradise), pp A112–A120. St. Lucia, Queensland: University of Queensland Press
21. Jenkinson D S (1982) The nitrogen cycle in long-term field experiments. Philosophical Transactions of the Royal Society (London) B296, 563–571

22. Kemp A, Hemkes O J and Steenbergen T van (1979) The crude protein production of grassland and the utilization by milking cows. Netherlands Journal of Agricultural Science 27, 36–47
23. Kolenbrander G J (1972) The eutrophication of surface water by agriculture and the urban population. Stikstof (English edn) 15, 56–67
24. Kolenbrander G J (1981) Leaching of nitrogen in agriculture. In Brogan J C, ed. Nitrogen losses and surface runoff from landspreading of manures, pp 199–216. The Hague, The Netherlands: Martinus Nijhoff/Dr Junk
25. LEI and CBS (1983) Landbouwcijfers 1983. The Hague/Voorburg, The Netherlands: Landbouw-Economisch Instituut and Centraal Bureau voor de Statistiek
26. Leusink A W F (1981) Onderzoek naar de hoeveelheid opneembare stikstof op grasland in het voorjaar. Regionaal Onderzoek Centrum Zegveld
27. Meer H G van der (1982) Effective use of nitrogen on grassland farms. In Corrall A J, ed. Efficient grassland farming. Proceedings of the 9th General Meeting of the European Grassland Federation (Reading), pp 61–68. Occasional Symposium No. 14 of the British Grassland Society. Hurley, UK: The British Grassland Society
28. Meer H G van der (1983) Efficiënt gebruik van stikstof uit kunstmest en organische mest op het weidebedrijf. NVWV, Gebundelde Verslagen No. 23, pp 67–80. Wageningen, The Netherlands: Nederlandse Vereniging voor Weide-en Voederbouw
29. Minderhoud J W (1960) Grasgroei en grondwaterstand. Proefstation voor de Akker- en Weidebouw, Publikatie No. 15
30. Minderhoud J W, Burg P F J van, Deinum B, Dirven J G P and Hart M L 't (1974) Effects of high levels of nitrogen fertilization and adequate utilization on grassland productivity and cattle performance, with special reference to permanent pastures in the temperate regions. Proceedings XIIth International Grassland Congress (Moscow), I 99–121
31. Morrison J, Jackson M V and Sparrow P E (1980) The response of perennial ryegrass to fertilizer nitrogen in relation to climate and soil: report of the joint ADAS/GRI grassland manuring trial – GM 20. Grassland Research Institute Technical Report No. 27. Hurley, UK: The Animal and Grassland Research Institute
32. Morrison J, Jackson M V and Williams T E (1974) Variation in the response of grass to fertilizer N in relation to environment. Proceedings No. 142, pp 28–38. London, UK: The Fertiliser Society
33. O'Connor K F (1974) Nitrogen in agrobiosystems and its environmental significance. New Zealand Agricultural Science 8, 137–148
34. Padmos L (1981) Aardgasprijs en stikstofprijs. Stikstof 98, 468
35. Penning de Vries F W T (1982) Systems analysis and models of crop growth. In Penning de Vries F W T and Laar H H van, eds. Simulation of plant growth and crop production, pp 9–19. Wageningen, The Netherlands: PUDOC
36. Prins W H (1980) Changes in quantity of mineral nitrogen in three grassland soils as affected by intensity of nitrogen fertilization. Fertilizer Research 1, 51–63
37. Prins W H (1983) Limits to nitrogen fertilizer on grassland. Doctoral thesis, Agricultural University, Wageningen
38. Prins W H, Burg P F J van and Wieling H (1980) The seasonal response of grassland to nitrogen at different intensities of nitrogen fertilization, with special reference to methods of response measurements. In Prins W H and Arnold G H, eds. The role of nitrogen in intensive grassland production. Proceedings of an international Meeting of the European Grassland Federation (Wageningen), pp 35–49. Wageningen, The Netherlands: PUDOC
39. Proefstation voor de Rundveehouderij (1977) Handboek voor de rundveehouderij. Tweede herziene druk. Lelystad, The Netherlands: PR
40. Proefstation voor de Rundveehouderij (1980) Handboek voor de rundveehouderij. Derde herziene druk. Lelystad, The Netherlands: PR
41. Richards I R and Hobson R D (1977) The relationship between nitrogen supply and the nitrogen yield of cut grass swards. The Journal of Agricultural Science (Cambridge) 88, 289–292

42. Ryden J C (1982) The nitrogen cycle in grassland – a case for studies in grazed pasture. The Grassland Research Institute, Essay Paper No. 2, Annual Report 1982, pp 150–166. Hurley, UK: The Animal and Grassland Research Institute
43. Ryden J C (1986) Gaseous losses of nitrogen from grassland (this volume)
44. Sibma L and Alberda Th (1980) The effect of cutting frequency and nitrogen fertilizer rates on dry matter production, nitrogen uptake and herbage nitrate content. Netherlands Journal of Agricultural Science 28, 243–251
45. Steenbergen T van (1977) Influence of type of soil and year on the effect of nitrogen fertilization on the yield of grassland. Stikstof (English edn) 20, 29–35
46. Stikstof (1972) Fertilizers and water quality. Stikstof (English edn) 15
47. Werkgroep 'Mineralen in krachtvoer in relatie tot bemesting en milieu' (1979) De gehalten van mengvoeders aan enkele minerale bestanddelen met betrekking tot de behoefte van de dieren, de uitscheiding in de mest en urine, alsmede enkele gevolgen voor bodem, plant en dier. Rapport No. 2. Wageningen, The Netherlands: Nationale Raad voor Landbouwkundig Onderzoek
48. Whitehead D C (1970) The role of nitrogen in grassland productivity. Commonwealth Bureau of Pastures and Field Crops, Bulletin 48. Hurley, UK: Commonwealth Agricultural Bureaux
49. Whitehead D C (1981) An improved chemical extraction method for predicting the supply of available soil nitrogen. Journal of the Science of Food and Agriculture 32, 359–365
50. Whitehead D C (1986) Sources and transformations of organic nitrogen in intensively managed grassland soils (this volume)
51. Whitehead D C, Barnes R J and Morrison J (1981) An investigation of analytical procedures for predicting soil nitrogen supply to grass in the field. Journal of the Science of Food and Agriculture 32, 211–218
52. Wieringa G W, Meer H G van der and Geurink J H (1980) Production and utilization of energy and protein in grass in relation to nitrogen rate, green crop fractionation and dairy feeding systems. In Prins W H and Arnold G H, eds. The role of nitrogen in intensive grassland production. Proceedings of an international Meeting of the European Grassland Federation (Wageningen), pp 103–111. Wageningen, The Netherlands: PUDOC
53. Wilkinson W B and Greene L A (1982) The water industry and the nitrogen cycle. Philosophical Transactions of the Royal Society (London), B296, 459–475
54. Wit C T de (1953) A physical theory of placement of fertilizers. Verslagen van Landbouwkundige Onderzoekingen 59.4
55. Woldendorp J, Dilz K and Kolenbrander G J (1965) The fate of fertilizer nitrogen on permanent grassland soils. In Burg P F J van and Arnold G H, eds. Nitrogen and grassland. Proceedings of the 1st General Meeting of the European Grassland Federation (Wageningen), pp 53–68. Wageningen, The Netherlands: PUDOC

The response to nitrogen fertilizer from a cut perennial ryegrass (*Lolium perenne* L.) pasture in the Scottish uplands relative to efficiency of fertilizer use and provision of herbage for animals

ANNE RANGELEY and PETER NEWBOULD
Hill Farming Research Organisation, Bush Estate, Penicuik, Midlothian EH26 0PY, Scotland

Key words: *Lolium perenne*, pasture, nitrogen fertilizer, nitrogen efficiency, nitrogen recovery, pasture utilization

Abstract. The dry matter (DM) responses by a perennial ryegrass pasture growing at a Scottish upland site to applications of 0–160 kg N ha^{-1} yr^{-1} in the sowing year and to 0–480 kg N ha^{-1} (given in three equal dressings) in the first harvest year were measured in a small plot cut experiment. The efficiency of use of N fertilizer was expressed in three ways: as the apparent N efficiency (kg DM per kg N); as the apparent N recovery (% N recovered of that applied) and as the amount of N fertilizer required to produce pasture to meet predicted animal requirements. The latter was based on a comparison of the dietary requirements of sheep calculated from knowledge of animal nutrition and grazing efficiency, with the quantities and patterns of herbage production found in the present experiment. Apparent N efficiencies ranged between 6 and 41 kg DM per kg N, and were influenced by growth stage, season of application and level of N and P fertilization. Apparent N recoveries ranged between 55 and 63% for the first harvest year and were greatly affected by season of N application. These results are compared with those from a similar experiment which was recently carried out at 21 sites in other parts of Britain. The herbage produced from pasture without N fertilizer would have supported about 6 ewes ha^{-1}. Applications of 120 and 240 kg N ha^{-1} yr^{-1} would have supported 10 and 15 ewes ha^{-1}, respectively, with only a few occasions when supply did not meet demand. It was difficult to match a set stocking rate with the herbage produced from 480 kg N ha^{-1} yr^{-1}, as depending on the animal numbers there would have been either major surpluses or deficits in herbage production at times during the grazing season. The use of N fertilizer to manipulate herbage production in upland sheep grazing systems is discussed.

Knowledge of the responses to level of application of nitrogen (N) fertilizer and especially of the efficiency of use of different N levels in contrasting sward types and on different soils throughout the growing season, is essential for the best use of pastoral resources by grazing animals. The amount of herbage available to the animal at any time and the characteristics of the sward e.g. height and botanical composition can be manipulated by strategic applications of N fertilizer. Low N availability from the soil is usually the most important factor limiting plant nutrition and growth in upland soils.

In a trial of upland sheep production systems at the Hill Farming Research Organisation (HFRO) [6], the N required to graze 10 or 15 ewes ha^{-1} was estimated empirically and was based on current recommendations by the East of Scotland College of Agriculture of 12 kg N ewe^{-1} yr^{-1} [8]. Application of 153 kg N ha^{-1} yr^{-1} in four unequal dressings during the season at both set stocking rates did not ensure

H G van der Meer, J C Ryden and G C Ennik (eds) Nitrogen Fluxes in Intensive Grassland systems.
 ISBN 978-94-010-8454-3

adequate amounts of herbage for the grazing sheep and there was great variability in response between fields and between plots in the same field.

A programme to investigate the reasons for the lack of consistency in herbage responses to applied N fertilizer in upland soils has started. The present paper describes a preliminary small plot study, carried out in 1982 and 1983, with herbage sampled by cutting to assess growth responses to N fertilizer and the efficiency of its use. The latter is defined in three ways: as the apparent efficiency of N fertilizer (ANE) which is expressed as the amount of harvested herbage dry matter (DM) produced per kg of N fertilizer applied; as the apparent recovery of N fertilizer (ANR) which is the amount of N present in the harvested herbage less the amount of N in the herbage of the unfertilized plots expressed as a percentage of that applied, and as the amount of N fertilizer required to produce sufficient herbage to meet the requirements of animals and maintain a well utilised sward.

Materials and methods

The experiment was set up at Hartwood Research Station, Lanarkshire, Scotland in the spring and summer of 1982. The site is at 200 m above sea level and the average rainfall is 1100 mm per year, with about 200 growing days where soil temperatures exceeded 6 °C. The soil is a non-calcareous gley of the Caprington series and on the experimental site itself, sandstone lay 32 cm below the soil surface.

The experiment consisted of 96 plots (2 m x 2 m) upon which 2 levels of lime, 4 levels of N and 4 levels of P fertilizer treatments were applied in fully factorial design with three replicates. Rates of application of magnesium limestone were 0 and 5 t ha^{-1} yr^{-1} and rates of N (as ammonium nitrate) and P (as superphosphate) were 0, 40, 80 and 160 kg element $ha^{-1} yr^{-1}$. The rate of application of K (as potassium chloride) to all treatments was 80 kg K $ha^{-1} yr^{-1}$. Unless otherwise stated, the information presented in this paper is only from the group of treatments which were given lime, 80 kg P ha^{-1} and the full range of N levels. The remaining results will be published separately.

A spray application of glyphosate was made on 30 April followed by rotavation and application of magnesium limestone on 13 May. The P, K and initial N were applied on 20 July and paraquat was sprayed onto the seedbed a week before sowing. Because of the need to reduce the weed seed population in the soil, the sowing date was 11 August which is somewhat later in the season than is normally recommended for pasture sown with a mixture containing white clover seed [2]. The seed mixture comprised of 25 kg ha^{-1} perennial ryegrass (*Lolium perenne* L.) cv. S23 and 5 kg ha^{-1} white clover (*Trifolium repens* L.) cv. Grasslands Huia. Despite the applications of herbicide before sowing, large numbers of *Ranunculus repens* L. seedlings appeared on the site during late August, so, on 26 September the site was sprayed with the selective herbicide Acumen (BASF, UK Ltd.). Herbage in the central square metre of each plot was first harvested on 21 October. The pasture was cut to a height of 3 cm, on this and all subsequent occasions.

In 1983, the first harvest year, N fertilizer was applied three times in the same form, at the same rates and to the same plots as the previous year. The dates of application were 5 May, 23 June and 1 September, resulting in annual application rates of 0, 120, 240 and 480 kg N ha^{-1}. The date of the first application of N was based on the attainment of a soil temperature of 5.5 °C at 10 cm [6]. Although in this season there was a warm spell in February during which the soil temperature exceeded 5.5 °C, it dropped below 5.5 °C during April and did not reach the threshold again until the beginning of May. Harvests were taken at approximately monthly intervals when the perennial ryegrass tillers had produced three leaves since the previous cut. The dates of harvest were 17 May, 20 June, 22 July, 29 August and 6 October.

In the experiment perennial ryegrass established well but white clover did not. This latter species contributed <5% of the dry matter produced from the pasture from sowing until August 1983 when it began to stolonate and spread, therefore the contribution to DM production during the whole season was small.

Harvested herbage was dried at 80 °C overnight and weighed. Subsamples were taken for Kjeldahl digestion and N analysis by the indophenol blue method [1]. Soil was collected for calculation of the available water capacity [9] and the rooting depth was noted at the same time.

The results were analysed statistically using analysis of variance; least significant differences between means (LSD) were calculated. The line which best fitted the response curve was estimated using the method of maximum likelihood.

Results

Dry matter responses to nitrogen fertilization

During the sowing year, responses to N fertilizer were small (Fig. 1), partly because the sowing date was late and also, perhaps, because the level of available N in the soil without N fertilizer was adequate for growth in the short late season. In 1983, the first harvest year, the treatment without N fertilizer produced 4.9 t DM ha^{-1} yr^{-1} and the

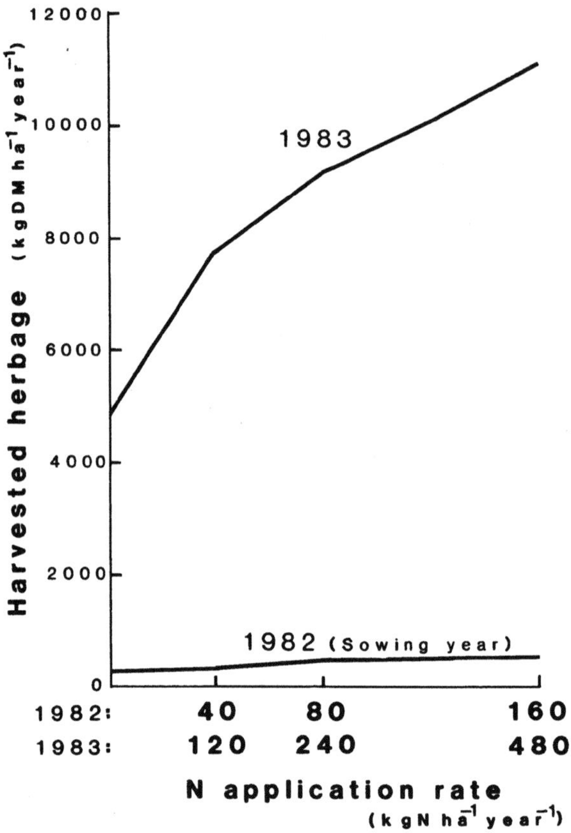

Figure 1. The response (kg DM ha^{-1} yr^{-1}) to fertilizer applications in 1982 and 1983.

response to N fertilizer was curvilinear. The maximum recorded production was 11.1 t DM ha^{-1} yr^{-1} for treatments given 480 kg N ha^{-1} yr^{-1} but the shape of the curve (Fig. 1) suggests that this was not necessarily the maximum amount of DM production. Small additional responses to higher dressing of N could have been expected. An exponential curve $Y = a + be^{-kx}$ best fitted the data ($a = 12\,272$, $b = 7357$, $k = 0.996$).

Pasture growth rates

On the whole, increasing levels of N fertilization significantly increased pasture growth rates (Table 1). Exceptions were during the first growth period when fertilizer dressings greater than 40 kg N ha^{-1} had no further effect and during the fourth growth period (22 July–29 August) when the growth rates by pasture given 40 and 80 kg N ha^{-1} were less than the control.

Table 1. Mean pasture growth rates (kg DM ha^{-1} day^{-1}) during 1983 after applications of N fertilizer

		Seasonal pasture growth rate (kg DM ha^{-1} day^{-1})				
Growth period	Date applied	N applied (kg ha^{-1})⁺ 0	40	80	160	LSD (P = 0.05)
1. 5–17 May	5 May	24	48	49	45	5.2
2. 17 May–20 June		48	86	100	135	10.2
3. 20 June–22 July	23 June	30	59	81	72	9.5
4. 22 July–29 Aug.		16	12	12	31	3.0
5. 29 Aug.–6 Oct.	1 Sept.	15	26	33	39	3.5

⁺Annual application rates were 3 × rate per dressing

The seasonal growth pattern approximately followed that expected for perennial ryegrass. However, growth was quite slow during the first growth period because it was cold (mean soil temperature at 10 cm was 4.7 °C) and during the fourth period because it was very dry (mean rainfall during that period was 0.77 mm day^{-1}). The mean soil temperature (10 cm) during the growing season was 8.1 °C and the mean rainfall was 2.3 mm day^{-1}.

Apparent efficiency of nitrogen fertilizer

There were large errors associated with the ANEs but the following trends were noted (Table 2). In the sowing year, DM production per

Table 2. The apparent efficiency of N fertilizer dressings (kg DM per kg N) showing the influence of growth phase, liming and P fertilization

	Apparent efficiency of N fertilizer (kg DM per kg N)			
	N applied (kg N ha^{-1})⁺ 40	80	160	LSD (P = 0.05)
Sowing year (1 cut)	1	2	2	2.2
First harvest year		Effect of season		
Spring N dressing (2 cuts)	40	26	21	
Summer N dressing (2 cuts)	19	18	12	10.9
Late summer N dressing (1 cut)	11	9	6	
Whole season	24	18	13	7.4
		Effect of lime*		
Unlimed (soil pH 5.3)	41	26	20	10.6
Limed (soil pH 5.8)	40	26	21	
		Effect of P fertilizer*		
P level (kg P ha^{-1})				
0	24	20	12	
40	26	29	18	15.0
80	40	26	21	
160	14	18	17	

⁺ Annual application rates were 3 × rate per dressing
*From spring dressing of N in first harvest year

Table 3. The apparent recoveries of N fertilizer dressings (as % of that applied) given on three occasions during the 1983 growing season

	Apparent recovery of N fertilizer (% of that applied)			
	N applied (kg ha^{-1})$^+$			LSD
	40	80	160	(P = 0.05)
First harvest year				
Spring dressing	116	93	106	
Summer dressing	21	39	52	37.9
Late summer dressing	36	33	31	
Whole season	58	55	63	25.1

$^+$Annual application rates were 3 × rate per dressing

kg N applied was low with ANEs between 1 and 2, but in the first harvest year they were greater, ranging between 6 and 41. Efficiencies were affected by level of application of N and by season. They decreased as the rate of N fertilizer increased and the later the date in the season that the dressings were applied.

Liming had no effect on ANEs (Table 2) although there was a DM response to this treatment of about 1 t $ha^{-1} yr^{-1}$ (data not shown). Phosphorus fertilization increased ANEs up to the application rate of 80 kg P $ha^{-1} yr^{-1}$, but with 160 kg P ha^{-1} the efficiencies decreased.

Apparent recovery of nitrogen fertilizer

As with the apparent efficiency of N fertilizer the errors associated with estimates of apparent recovery were high (Table 3). ANRs were greatest for the spring dressings of N in 1983. In fact this method of calculation resulted in apparent recoveries of more than 100%. The recoveries of the summer dressings of N increased as the rate of application increased. The low recoveries from the second dressing of 40 and 80 kg N ha^{-1} arose from the fact that at the fourth cut the fertilized pasture produced less DM than the control. Only one-third of the N applied in September was apparently recovered and, over the whole season 55 to 63% of fertilizer N was apparently recovered in the harvested herbage.

Soil nitrogen

The N yield from the control plots was 119 kg N ha^{-1} for the whole of the 1983 season. There was 28 kg N ha^{-1} in white clover herbage during the season and if it is assumed that all of this was from N_2 fixation then the N yield from the soil was about 91 kg N $ha^{-1} yr^{-1}$.

Discussion

Three parameters are used to characterise the N response: N_S, the amount of N in the herbage from treatments without N fertilizer; Y10, the yield at which the marginal response fell below 10 kg DM per kg N, i.e. when $\frac{dy}{dx} = 10$ on the response curve (that is about the limit which presently makes N application economically worthwhile [7]), and the ANE at the 300 kg N ha^{-1} yr^{-1} application rate. These parameters have been calculated previously for data from a series of trials in England and Wales known as GM20 [7].

The N_S from the control plots was about 91 kg N ha^{-1} for the season and is relatively high when compared to the overall mean of 60 kg N ha^{-1} for GM20. At the Hartwood site the previous crop was permanent pasture and in the sowing year the soil had been ploughed and rotavated and herbicide was applied to kill weeds. These treatments may all have stimulated mineralisation [4].

Some of the variation in the annual DM produced in GM20 was explained by an equation relating Y10 to the amount of water available to the plants during the growing season. When data for the water available to pasture at Hartwood during 1983 were used in this equation the calculated result for Y10 was 8.34 t DM ha^{-1} yr^{-1} and the Y10 from the exponential equation which described the response curve was 9.6 t DM ha^{-1} yr^{-1}.

At Hartwood the ANE from applications of 300 kg N ha^{-1} yr^{-1} was about 16 kg DM per kg N and ANRs for the whole season ranged between 55 and 63%, values somewhat less than the corresponding means found in GM20. Drought in August 1983 which depressed growth rates, and the slow growth rates and the presence of clover in the zero N treatments in September all contributed to the lower recoveries and efficiencies of N fertilizer.

After comparing the data from Hartwood and the GM20 experimental sites, it is considered that there are no unusual features about the response to N fertilizer at Hartwood in 1983. Therefore, further consideration of N fertilizer efficiency is directed towards the objective of providing sufficient DM to meet animal requirements and maintain a well utilised sward.

The objective of upland farming systems is to maximise the economic return and to optimise the inputs required for herbage production to provide both for grazing during the pasture growing season and for conservation for winter feed. This is being attempted at Hartwood by controlling pasture mass or height to maximise the growth rate and minimise senescence while at the same time offering pasture of an amount and quality to attain maximum output of lamb [6]. Part of the grazed area is fenced off for conservation during June and July when pasture growth rates exceed animal requirements. The actual area depends on the particular circumstances and so adds flexibility to the system.

Recent experiments on management of grazed pastures have shown that a mass of 1250 kg DM ha^{-1} with a leaf area index of 3–4 and a sward height of 3–4 cm meets the requirements described above [3]. However, maintenance of a sward with this height throughout the season is constrained by slow pasture growth rates during cold and dry weather and the need to accumulate herbage in autumn for consumption during the mating period in November when pasture growth has stopped. Therefore a target seasonal profile of pasture heights has been suggested which starts in spring at about 2 cm and increases gradually during summer to about 7 cm in late September (T.J. Maxwell, unpublished).

The pasture and winter feed requirements of ewes during an annual cycle can be assessed according to the number of lambs they carry. So the pasture growth rates required to feed the animals and maintain the season height profile can be calculated. N fertilizer is the main tool available to increase pasture growth rate to meet increases in stocking rates or in nutritional demand.

Although the present study was conducted using intermittently cut pasture which may not precisely represent growth rates under grazing, relationships between increases in growth rate due to N fertilization and probable animal carrying capacities have been calculated. Herbage profiles estimated to be needed to provide sufficient pasture for cross bred ewes at 2 stocking rates (10 and 15 ewes ha^{-1}) and carrying 1.5 lambs per sheep as used in the Hartwood Systems trial are shown in Fig. 2 together with the pasture growth rates and N responses found to have the best fit. Despite the fact that growth rates did not meet animal needs in spring and August (possibly due to late application of N and/or drought) the requirements of the pasture are broadly met by 120 and 240 kg N ha^{-1} yr^{-1} for the two stocking rates, respectively, applied in 3 dressings. These rates are equivalent to 12 or 16 kg N per ewe, the latter rate being considerably higher than that recommended [8] despite the relatively high rate of mineralisation in this soil.

In the present experiment in the absence of N fertilizer the carrying capacity of the pasture would have been about 6 ewes ha^{-1}. Matching a stocking rate to the highest level of fertilizer application proved difficult. With 480 kg ha^{-1} yr^{-1}, depending on the time of year, the stocking rates required to consume the herbage varied from 15 to 33 ewes ha^{-1}. The lower rates would lead to surpluses of herbage in summer and accumulation of dead poor quality material. The higher rates would result in deficiencies of herbage at times of the year when pasture

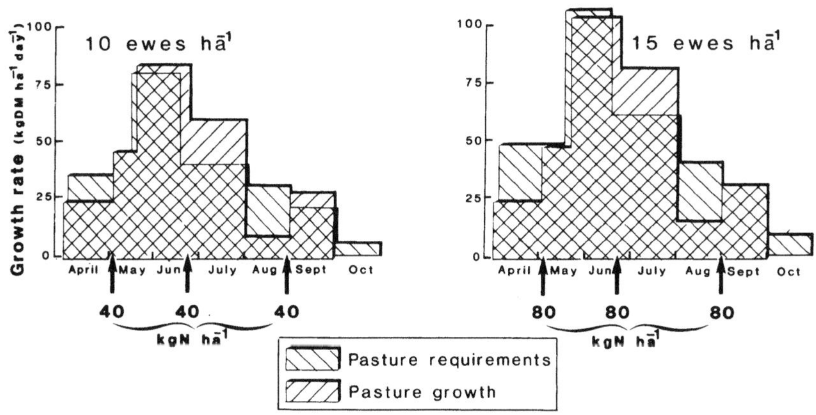

Figure 2. N fertilizer needed to attain the required pasture growth rates for a Greyface sheep system at Hartwood from a cut pasture in 1983.

growth rate was slow and controlled by factors other than N fertilizer e.g. moisture or temperature. Under the latter circumstances either the body condition of the animals would decline or expensive supplementary feeding would be required. Clearly, in any farming system decisions on levels of N fertilization and stocking are influenced by the availability of alternative feed resources. Further refinement of the pattern and rate of application of N fertilizer to grazed swards may alter the seasonal growth rate curves [5] and lead to better and more economical matches between pasture supply and animal utilization.

A recent survey of sown grassland in the East of Scotland showed that the average amount of N fertilizer used was about 100 kg N $ha^{-1} yr^{-1}$ and application rates to 40% of the farms surveyed were 60 kg N $ha^{-1} yr^{-1}$ or less. Average stocking rates were 1.7 liveweight units per ha or the equivalent of 0.43 lowground ewes and lambs per ha grazed for 12 months [10]. This information when compared with the results from the present study suggests that pastures in this region of Scotland are grossly under-utilized.

Acknowledgements. The authors would like to thank Mr G.R. Bolton for help in conducting the experiment.

References

1. Allen S E (1974) Chemical analysis of ecological materials, pp 449–452. Oxford, UK: Blackwell
2. Anon (1983) White clover. Publication 99. The Scottish Agricultural Colleges
3. Bircham J S and Hodgson J (1983) The influence of sward conditions on rates of herbage growth and senescence in mixed swards under continuous grazing management. Grass and Forage Science 38, 323–331

4. Campbell C A (1978) Soil organic carbon, nitrogen and fertility. In Schitzer M and Khan S E, eds. Soil organic matter. Developments in Soil Science 8, pp 95–136. Amsterdam: Elsevier
5. Dawson K P, Jewiss O R and Morrison J (1983) The effects of strategic application of nitrogen on the improvement of seasonal yield from perennial ryegrass swards and its interaction with available water. In Corrall A J, ed. Proceedings of the 9th General Meeting of the European Grassland Federation, Reading, 1982, pp 171–174. Hurley, UK: The British Grassland Society
6. Maxwell T J (1983) Factors affecting the growth and utilization of sown grasslands for sheep production. In Haresign W, ed. Sheep production. 35th Easter School in Agricultural Science, Nottingham, pp 187–205. London, UK: Butterworth
7. Morrison J, Jackson M V and Sparrow P E (1980) The response of perennial ryegrass to fertilizer nitrogen in relation to climate and soil. Technical Report 27, Hurley, UK: The Grassland Research Institute
8. Rutter W (1982) Sheep from grass. Bulletin 13. Edinburgh, UK: East of Scotland College of Agriculture
9. Salter P J and Williams J B (1967) The influence of texture on the moisture characteristics of soils. IV. A method of estimating the available-water capacities of profiles in the field. Journal of Soil Science 18, 174–181
10. Swift G, Holmes J C, Cleland A T, Fortune D and Wood J (1983) The grassland of East Scotland – A survey 1976–78. Bulletin 29. Edinburgh, UK: East of Scotland College of Agriculture

Efficient use of fertilizer nitrogen on grass swards: effects of timing, cutting management and secondary grasses

R D HARKESS and J FRAME
The West of Scotland Agricultural College, Auchincruive, Ayr KA6 5HW, Scotland

Key words: efficient nitrogen use, nitrogen timing, cutting systems, secondary grasses

Abstract. Efficient use of nitrogen (N) is necessary for cost effective grass production and to minimise risks of water pollution. A guide to spring N application, the Dutch T sum 200°, has been compared with two accumulated soil temperature systems. When used in conjunction with soil and weather conditions, all the systems are of practical value. The temperature targets require adjustment for altitude, and allow a degree of flexibility in timing of N application. Potential herbage yield and quality from a series of silage cutting systems were assessed. By manipulating the date of first cut, level of N, regrowth interval and number of cuts, it was possible to devise multi-cutting systems which would attain specified yield and quality targets for various livestock systems. Certain grasses common in permanent grassland e.g. *Holcus*, *Festuca* and *Agrostis* species, can outyield *Lolium perenne* at N levels up to 120 kg N ha^{-1} yr^{-1} but not above. However, the digestibility of *L. perenne* is superior to the other species. Several secondary grasses had high N concentrations because they did not metabolize high N uptakes into high dry matter yields. In certain situations, better management of the existing species may be preferable to reseeding.

In intensive grassland farming, there is an increasing need for efficient use of fertilizer nitrogen (N) on three counts, namely, environmental considerations, possible input savings, and economic response of swards to applied N. Many factors other than soil type, soil fertility and climate influence efficient use of N. These include the frequency and severity of defoliation and the type of grassland. Three aspects of current work on efficient use of N in the west of Scotland are presented.

There is much interest in the UK [2, 6, 13] in the Dutch T sum 200° technique [11, 12] for predicting the best time to apply fertilizer N in the spring. In west Scotland, application of spring N also has been associated with the accumulated 100 mm soil temperature (at 0900 h) from 1 February, a system referred to as the T value 100° [14]. Also, N application has been linked to the accumulated 300 mm soil temperature (from 1 February) and referred to as the T soil 150° [10]. The single daily temperature reading and the shorter data collection period for the soil systems offer some practical advantages for on-farm recording by farmers.

The yield versus quality controversy for silage production has received considerable attention in recent years [7, 8, 9]. Debate has been encouraged by the call for better quality winter fodder. Yield and quality are particularly influenced by rate of N application, the date of

H G van der Meer, J C Ryden and G C Ennik (eds) Nitrogen Fluxes in Intensive Grassland systems.
 ISBN 978-94-010-8454-3

the first harvest and by the frequency with which regrowths are harvested. The concern of farmers is that improved quality will lead to lower yields and the inability to make sufficient winter feed. Work reported here has examined different cutting regimes which incorporate various levels of N application in order to define fertilizer/cutting systems for specific animal enterprises.

Although indigenous or 'secondary' grasses such as *Agrostis*, *Festuca* and *Holcus* species are widespread in British grasslands, there is a lack of data on their yield, quality and response to differing rates of application of fertilizer N [3, 5]. A series of experiments has been started to assess the productivity of these grasses and the results of the first three years are presented.

Materials and methods

Experimental sites

The spring N trial plots were sited on established grass swards at Ayr, 45 m above sea level (asl) and on a farm at Lanark, 200 m asl (Latitude 55° 28′/40′ N). The soils were silty clay loams of pH 6.08 and 5.85, respectively, and classed as freely to imperfectly drained. Swards were 5–7 years old and comprised mainly *Lolium perenne* with some *Phleum pratense* and *Trifolium repens*.

The cutting management and secondary grasses trials were on plots 2–4 years old sown at 15 m asl at Ayr on a sandy loam with a pH 6.2. Grasses sown are discussed below but included species and cultivars of *Lolium*, *Holcus*, *Festuca*, *Agrostis* and *Poa*.

Fertilizer application

Spring N trials received 90 kg N ha^{-1} as ammonium nitrate along with 26 kg P ha^{-1} and 50 kg K ha^{-1} as a 0:19:19 compound fertilizer. At least six application dates were spread out to cover the target temperatures viz, Dutch T sum 200°, T value 100° and T soil 150°. First application dates ranged from 5 February to 12 March at the lowland site and from 6 February to 2 March at the upland site.

Over the trial period, the January to April mean daily 100 mm and 300 mm soil temperatures were 3.9 and 5.0°C, respectively, at the lowland site; at the upland site temperatures were one degree lower. The mean daily air temperature was 5.0 and 2.8°C for the two sites and rainfall was *circa* 245 mm.

In the cutting management trials, a 20:6:12 compound fertilizer was applied to give three levels of N application viz, 100 kg N ha^{-1} for the first cut followed by 2 kg N ha^{-1} day^{-1} for regrowths, 120 kg N ha^{-1}

Table 1. Accumulated temperature for best dry matter yields using three prediction systems (averages 1979–83)

		Site 1 (45 m asl)		Site 2 (200 m asl)	
Prediction system	Temperature recorded	Accumulated temperature for best yield	Date of best accumulated temperature	Accumulated temperature for best yield	Date of best accumulated temperature
		1979–83		1980–83	
Dutch T sum 200°	Air, (max + min)/2	236°	3 March	246°	21 March
T value 100°	Soil, 100 mm	75°	1 March	122°	20 March
T soil 150°	Soil, 300 mm	102°	28 February	153°	20 March

followed by 2.5 kg N $ha^{-1} day^{-1}$ and 140 kg N ha^{-1} followed by 3.0 kg N $ha^{-1} day^{-1}$. Total N input was 260, 318 and 378 kg N $ha^{-1} yr^{-1}$. In a second series of trials fertilizer N was applied at rates ranging from 150 to 450 kg N $ha^{-1} yr^{-1}$ and management designed to simulate different combinations of silage cuts and grazings as discussed below.

In the studies involving secondary grasses, 0, 120, 240, 360 and 480 kg N $ha^{-1} yr^{-1}$ were applied as a 20:6:12 compound fertilizer in six equal amounts at intervals of four weeks from early March.

Harvesting

N-timing trials were cut with an Agria motor mower, the grass from a 5 m × 1.2 m area being weighed and sub-sampled for determination of dry matter (DM), N and digestible organic matter content. The spring N plots were cut when 1.5 to 2.0 t DM per ha was available on the most advanced plots.

In the cutting frequency trials, silage cuts were taken between late May and mid September to simulate a range of on-farm cutting times, later grass growth being sampled at a frequency to simulate autumn grazings. The secondary grasses were cut six times to simulate six grazings.

Results and discussion

Efficient use of nitrogen in spring

The response to spring N was *circa* 14 kg DM per kg N applied (range, 10–18). Table 1 shows the accumulated temperatures which gave the best yields between 1979 and 1983 for seven trials at the lowland site and three at the upland site.

The Dutch T sum 200° prediction was 10–14 days early. The suggested target accumulated soil temperatures for the T value 100°

and T soil 150° systems were 7–14 days late at the lowland site but were early at the upland site. Site differences of this type are in accord with the findings of Jagtenberg [11]. The dates at which the systems reached best accumulated temperatures were very close as noted in Table 1. These dates are at least two weeks earlier than traditional spring N application dates. The change in accumulated temperature was approximately 25° weekly regardless of system or site, so an 'early warning' for lowland or early sites at Dutch T sum 200°, T value 50° or T soil 75° would alert farmers on the approach of suitable conditions. Indeed, if soil conditions and the weather forecasts were good, early fertilizer N could be applied at these times. This work will continue with particular reference to upland sites since about two-thirds of the grassland in Scotland is above 125 m.

Concern about inefficient use due to leaching or denitrification losses in cool wet spring weather is given as a reason for farmers being late in applying fertilizer N in spring. The data for the lowland site indicate that production can be lost at an important time of year. In the short growing season in northern Europe it is vital that timely application of fertilizer be made. The Dutch T sum 200° system has given a reasonable result. However, the target temperatures of the soil systems may have to be corrected to lower values (e.g., T value 80° or T soil 100°) for the lowland site (i.e., below 100 m asl) and increased for upland sites once further information becomes available. A similar trend has been recorded for sites in east and north Scotland [15].

A national newspaper publishes a weekly update on Dutch T sum 200° for the UK [4]. However, the 50 km × 50 km grid is too large for the prediction to be accurate due to variations in soil type, degree and direction of slope (aspect) and altitude among farms in the UK. Hence, interest in individual farm recording has increased.

Cutting management

From 1977 to 1979 a 3-year trial has aimed to produce at least 10 t DM $ha^{-1} yr^{-1}$ at D value 70 (metabolisable energy = 11.2 MJ per kg DM) from repeated harvests from the same sward of *L. perenne* cv. Animo (D value = % digestible organic matter in the DM). The mean annual yields obtained for the three rates of N application (260, 318, 378 kg $ha^{-1} yr^{-1}$) were 8.2, 8.6 and 8.8 t DM ha^{-1}, respectively [8]. Responses to the incremental amounts of N above the lowest level applied (260 kg N ha^{-1}) were only 7 and 3 kg DM per kg N. Where regrowth intervals of 3, 4 or 5 weeks followed the first cuts, mean annual total yields of 8.5, 8.8 and 8.2 t DM ha^{-1} were obtained.

The best overall system in this trial was when the first cut on 8 June was followed by three harvests at monthly intervals with an N input of

Table 2. Yield and quality of herbage from three silage and grazing systems each at three rates of N application (*Lolium perenne* cvs Premo + Morenne) 1982–83

System	Rate of N (kg ha^{-1} yr^{-1})	DM (t ha^{-1} yr^{-1})				S + G yield response to N[c]		Weighted D value – Silage	
		1982		1983					
		S[a]	S + G[b]	S[a]	S + G[b]	1982	1983	1982	1983
(A) *2 Silage cuts* (+ 3 grazings) – first cut 7 June; subsequent cuts at intervals of 50 days									
	150	9.6	10.0	7.3	8.7	–	–	68.9	66.6
	250	11.8	13.5	9.4	11.3	35	26	66.3	65.9
	350	13.8	16.0	10.6	13.1	25	18	64.0	65.3
(B) *3 Silage cuts* (+ 1 grazing) – first cut 28 May; subsequent cuts at intervals of 50 days									
	250	12.1	12.6	11.1	12.1	–	–	68.9	69.0
	350	13.5	14.3	11.6	12.8	17	7	68.5	67.8
	450	14.2	15.2	12.2	13.4	9	6	67.9	66.3
(C) *4 Silage cuts* (+ 1 grazing) – first cut 18 May; subsequent cuts at intervals of 40 days									
	250	10.2	10.6	8.9	9.8	–	–	72.7	71.5
	350	11.4	12.0	10.6	11.9	14	21	71.4	72.3
	450	13.7	14.4	11.0	12.4	24	5	71.5	70.0

a = Silage
b = Silage + Grazing
c = kg DM per kg N

120 kg ha^{-1} for the first cut plus 2.5 kg per ha per day regrowth following each cut. Over all regrowth cuts for the three years, herbage of 70 D value was obtained at only 25, 33 and 11% of the harvests for the 3, 4 and 5 week regrowths respectively. If the quality target was lowered to 68 D then 69, 69 and 57% of all the regrowths reached this value.

A second series of trials examined five cutting systems at each of three rates of N application on two sward types. Table 2 shows results for three systems for one of the swards. The intensive frequent cutting (System C) was successful in achieving a high quality silage; D values were consistently over 70. However, this system could not sustain the silage or annual yields of the more traditional System A without additional fertilizer N. On-farm consideration must weigh the cost and quality of production against the cost of alternative feeds. In Systems B and C, designed to produce high quality silage (68–72 D value), DM response to N was reduced in 1983 due to dry summer weather. System B (3 cuts) increased silage yield and quality over System A (2 cuts). System C (4 cuts), whilst improving silage quality, was lower yielding than System A at similar N levels. Silage quality target is determined by the needs of the livestock system and the ability to make sufficient silage. Data such as those in Table 2 will assist with stocking rate/feeding decisions as well as providing information for computer modelling studies.

Secondary grasses

Response of 12 grasses, sown in monoculture, to five rates of fertilizer N applied in six equal amounts at monthly intervals was evaluated in another series of studies. Table 3 presents selected data for six of the grasses; data for the other six are available elsewhere [5]. *Holcus lanatus*, *Festuca rubra* and *Agrostis stolonifera* were superior to *L. perenne* in DM yield at low to moderate rates of N application (0 and 120 kg $ha^{-1} yr^{-1}$) but this superiority was not sustained as N rate increased. *H. lanatus*, *F. rubra* and *Poa pratensis* were the highest yielding secondary grasses at high rates of N application.

The superiority of *L. perenne* in D value was unchallenged by the secondary grasses at all rates of N. *F. rubra*, *Agrostis* species and *P. pratensis* formed a group of 'low D value' grasses. *L. perenne* had consistently low N contents at all rates of N and hence low crude protein yields except at the highest rate of N application. Conversely, *Agrostis* species had a high N content, reflecting a high yield of crude protein, a response similar to that observed for *Phleum bertolonii* and *Poa trivialis* [5].

Yield data reflected the superiority of *L. perenne* in its response to N (Table 3). *F. rubra* and *Agrostis* species (Table 3) and *P. trivialis*

Table 3. Performance of six grass species at five rates of N application. 1979–81

Species	Applied N (kg ha^{-1})	DM (t ha^{-1})	D value	N (%)	Response to N[a]	Apparent N recovery[b]
Lolium	0	2.39	72	1.95		
perenne	120	6.13	73	1.94	31.2	60
(cv Perma)	240	9.72	72	2.11	29.9	66
Source: NL	360	11.95	71	2.27	18.6	62
	480	13.21	71	2.70	10.5	65
Holcus	0	3.69	68	2.11		
lanatus	120	6.69	68	2.22	25.0	59
(commercial)	240	8.80	68	2.56	17.6	57
Source: FRG	360	10.57	68	2.58	14.8	54
	480	11.47	69	2.98	7.5	55
Festuca	0	3.28	65	1.92		
rubra	120	6.36	65	2.03	25.7	55
(cv Aberystwyth S 59)	240	9.00	64	2.29	22.0	59
Source: UK	360	11.05	64	2.50	17.1	59
	480	10.87	65	2.85	−1.5	52
Agrostis	0	2.64	65	2.21		
stolonifera	120	6.30	64	2.38	30.5	79
(commercial)	240	8.55	64	2.66	18.8	70
Source: USA	360	10.06	64	2.80	12.6	62
	480	10.49	64	3.23	3.6	59
Agrostis	0	2.23	66	2.24		
castellana	120	5.37	67	2.43	26.2	68
(cv Highland)	240	8.76	66	2.83	28.3	83
Source: USA	360	10.10	65	2.98	11.2	70
	480	10.65	66	3.41	4.6	65
Poa	0	1.50	66	2.24		
pratensis	120	4.84	64	2.38	27.8	71
(cv Newport)	240	8.04	64	2.58	26.7	73
Source: USA	360	10.17	64	2.75	17.8	68
	480	11.76	64	3.02	13.3	67

[a] *kg DM per kg N applied for each increment*
[b] *increase in N yield over control N yield (N_0) as % of total N applied*

and *Anthoxanthum odoratum* [5] gave a poor response at the upper N increment. Other species were intermediate in response, although *P. pratensis* (Table 3) and *Cynosurus cristatus* [5] were very good at the final N increment. Apparent N recovery was generally high at the various N rates for *Agrostis* species and *P. pratensis* (Table 3) and *P. bertolonii* [5], but these species were unable to metabolize high N uptakes into high DM yields.

Conclusions

The use by advisers of accumulated temperature systems as a guide to the time of application of fertilizer N in spring has stimulated much farmer interest in the UK. Those who have applied early N have been sufficiently encouraged to advance their normal spring N application date by 10 to 20 days [15]. The technique is not likely to become an exact procedure and there is some spread of optimum dates before and after the target accumulated temperature [10]. This flexibility has a practical value since it gives the farmer room to manoeuvre according to soil and weather conditions. Some degree of annual variation in accuracy does occur [1]. Nevertheless, when used in conjunction with weather reports and in accord with farm management objectives, the technique may offer a useful guide to the more efficient use of early N on the farm.

Many of the past experiments reported in the literature concerning frequency of cutting, sward production and herbage quality were not devised with practical grassland management systems in mind. Recent trials at this Institute were designed to overcome these limitations, and the findings indicate: (i) the importance of date of first cut in influencing annual silage yield; (ii) the need to increase the number of cuts to sustain yields where high quality silage is required; (iii) the difficulty of sustaining a high D value from successive regrowth harvests; (iv) the falling DM response to additional N where N inputs are already high; (v) the value of high rates of N in sustaining yields under frequent cutting management. The data suggest that there is scope on many farms to improve the integration of silage and grazing into effective systems which make optimum use of fertilizer N in meeting production and quality targets.

Secondary grasses have been largely condemned in the past since they are associated with poor soil fertility and consequent low production. The work reported here and elsewhere [5] suggests that several secondary grasses can match or outyield *L. perenne* at low to moderate rates of N application. Therefore, unless high N regimes are to be used, it may be unnecessary to replace secondary grasses by sown species. However, where high production and superior herbage quality are required in conjunction with effective N use, the value of *L. perenne* is unquestioned. Clearly, elucidation of the merits and demerits of secondary grasses is warranted in order to make decisions about reseeding and to guide farmers on intensity of N use for permanent grassland.

References

1. Burg P F J van (1977) Nitrogen fertilization of grassland. In Gilsenan B, ed. Animal production from temperature grassland. Proceedings of an international Meeting, Dublin, pp 104–108
2. Daly M and MacKenzie G H (1983) The timing of spring N application to grass. Grass and Forage Science 38, 149

3. Dibb C and Haggar R J (1979) Evidence of effect of sward changes on yield. In Charles A H and Haggar R J, eds. Changes in sward composition and productivity. Occasional Symposium No 10 of the British Grassland Society, 1978, pp 11–20. Hurley, UK: The British Grassland Society
4. Farmers Weekly (1983) T sum totals. Weekly newspapers 21 January–18 March 1983
5. Frame J (1983) Yield and quality response of secondary grasses to fertilizer nitrogen. In Corrall A J, ed. Efficient grassland farming. Proceedings of the 9th General Meeting of the European Grassland Federation, Reading, 1982, pp 292–294. Occasional Symposium No 14 of the British Grassland Society. Hurley, UK: The British Grassland Society
6. Garstang J R (1980) The date of nitrogen application to upland pasture in the spring in a high-rainfall maritime climate. In Prins W H and Arnold G H, eds. The role of nitrogen in intensive grassland production. Proceedings of an international Symposium of the European Grassland Federation, p 159. Wageningen: PUDOC
7. Harkess R D (1981) Frequency of cutting and herbage quality for silage. The West of Scotland Agricultural College, Agronomy Department, Publication 627
8. Harkess R D (1983) Aspects of cutting strategy on herbage production and quality. In Corrall A J, ed. Efficient grassland farming. Proceedings of the 9th General Meeting of the European Grassland Federation, Reading, 1982, pp 305–306. Occasional Symposium No 14 of the British Grassland Society. Hurley, UK: The British Grassland Society
9. Harkess R D (1983) Cutting management and sward productivity. The West of Scotland Agricultural College, Agronomy Department, Publication 760
10. Harkess R D (1983) Spring grass response to fertilizer N applied according to cumulative temperature systems. The West of Scotland Agricultural College, Agronomy Department, Publication 750
11. Jagtenberg W D (1970) Predicting the best times to apply N to grassland in spring. Journal of the British Grassland Society 24, 266–271
12. Postmus J and Schepers J H (1980) Temperature sum and date of spring application of nitrogen on grassland – results in the Netherlands. In Prins W H and Arnold G H, eds. The role of nitrogen in intensive grassland production. Proceedings of an international Symposium of the European Grassland Federation, p 159. Wageningen: PUDOC
13. Sandford H (1980) Temperature sum and date of spring application of nitrogen on grassland – results in the United Kingdom. In Prins W H and Arnold G H, eds. The role of nitrogen in intensive grassland production. Proceedings of an international Symposium of the European Grassland Federation, p 160. Wageningen: PUDOC
14. Swift G (1983) A prediction system for timing of N for spring grass. In Corrall A J, ed. Efficient grassland farming. Proceedings of the 9th General Meeting of the European Grassland Federation, Reading, 1982, pp 331–332. Occasional Symposium No 14 of the British Grassland Society. Hurley, UK: The British Grassland Society
15. Swift G, Mackie C K, Harkess R D and Franklin M F (1985) Time of fertilizer nitrogen for spring grass. The Scottish Agricultural Colleges, Research and Development Note 24

Nitrogen supply and the persistence of grasses

T J BEHAEGHE
Rijksuniversiteit, Coupure links 653, B-9000 Gent, Belgium

Key words: persistence, growth cycle, nitrogen application, cutting date, *Lolium perenne*, *Elymus repens*

Abstract. After a period of vigorous growth, hibernated grass tillers are killed in the late spring and replaced by a new generation. This transition of tiller generations can be gradual and overlapping, or abrupt thereby weakening the persistence of the grass. Conditions of a heavy cut following high nitrogen fertilization in spring are especially weakening, sometimes killing whole tufts of *Lolium perenne*. This is followed by invasion of adventive species such as *Poa annua*, *Stellaria media* and ultimately *Elymus repens*.

Pasture grasses are 'perennial' by a succession of tiller generations [3, 4]. Unlike other perennial plants, however, the shoots of which are viable from spring to autumn, the tillers of pasture grasses have a life cycle from June to June (Fig. 1). Hence, spring growth arises from hibernated tillers and summer growth develops from a new generation of tillers.

The summer tillers continuously produce new leaves from a protected and winter-hard apex. The production of new leaves would continue for more than one year, if there was no hibernation, as summer grass is morphologically and physiologically the persistent form. Hibernation, however, profoundly changes this condition, inducing a high growth rate, flower initiation and stem elongation [1, 2, 3]. Seed production would result in the natural death of the tiller, but defoliation generally

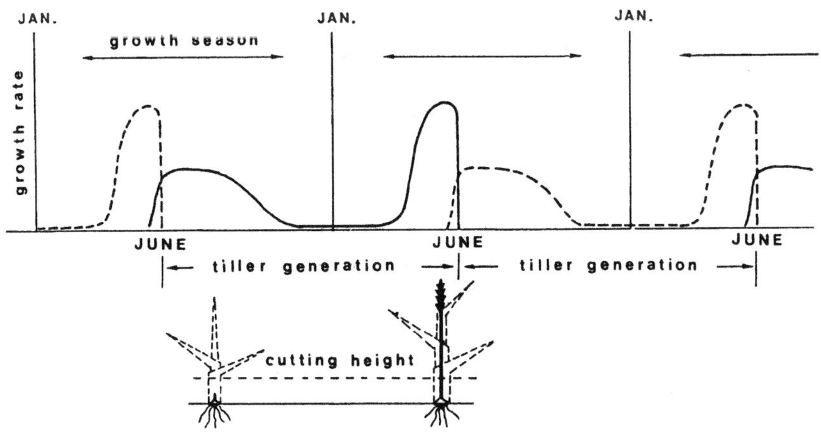

Figure 1. Schematic representation of the growth and development of pasture grass.

H G van der Meer, J C Ryden and G C Ennik (eds) Nitrogen Fluxes in Intensive Grassland systems.
 ISBN 978-94-010-8454-3

precipitates the death by decapitating the apices of the elongating tillers, mostly before heading.

Transition of tiller generations and midsummer depression of growth

As a general rule, only elongated tillers will be killed by defoliation. In spring or early summer most main tillers are elongating, and in fact, are killed by a defoliation following that elongation. When and how many tillers are killed, depends greatly on (i) whether the sward is under grazing or cutting management, (ii) the date of defoliation and (iii) the number of elongated tillers as influenced by season, variety, rate of nitrogen (N) application and management. Differences in the degree of tiller mortality between species and cultivars are illustrated by data in Table 1, between cutting and grazing by data in Table 2 and between early and late cutting by data in Table 3.

Some conditions result in almost 100% mortality, such as those prior to the second cut of *Phleum pratense* (Table 1). In such cases, the

Table 1. Tiller mortality and growth rate as influenced by cultivar and species. Successive cuts, 400 kg N ha^{-1} yr^{-1}

	% mortality after cutting on						
	5.5	5.29	6.27	7.22	8.20	9.16	10.20
L. perenne 'Vigor'	1	42	32	6	12	1	0
L. perenne 'Melino'	25	87	24	6	6	0	1
Phl. prat. 'Erecta'	3	100	17	65	35	4	0

	growth rate between cuts, kg DM ha^{-1} day^{-1}					
L. perenne 'Vigor'	115	105	81	63^a	58	41
L. perenne 'Melino'	119	79	74	58^a	54	33
Phl. prat. 'Erecta'	145	44	74	42^a	41	25

a = drought period

Table 2. Tiller mortality and growth rate as influenced by successive cutting or grazing of *Lolium perenne* cv. Vigor. N rate under cutting: 90 kg N ha^{-1} in spring, followed by 80 kg N ha^{-1} cut^{-1}. N rate under grazing: 90 kg N ha^{-1} in spring, followed by decreasing rates to 42.5 kg N ha^{-1} before the last grazing

	% mortality due to defoliation on following dates					
	5.15	6.7	7.6	8.2	9.10	10.12
cutting	16	63	20	4	2	2
grazing	0	14	17	4	3	0

	growth rate between defoliations, kg DM ha^{-1} day^{-1}				
cutting	117	95	100	26^a	55
grazing	125	144	99	46^a	52

a = drought period

Table 3. Tiller density, tiller mortality and regrowth in a sward of *Lolium perenne* cv. Vigor as influenced by the rate of application of fertilizer N and the date of first cut. Herbage was removed immediately after cutting

	1st cut on 8 May (early cut)			1st cut on 28 May (late cut)		
Rate of N application, $kg\,N\,ha^{-1}$ on 6 March 1984	60	110	160	75	137.5	200
Tillers per dm^2 after 1st cut:						
(1) living main tillers	87	86	93	53	44	3
(2) living lateral tillers	128	153	156	71	87	42
(3) decapitated main tillers	2	4	14	18	70	73
total	217	243	263	142	201	118
Tillers decapitated:						
per cent of total	1	2	5	12	35	62
per cent of main tillers (1 + 3)	2	4	13	25	61	96
DM yield at 1st cut ($kg\,ha^{-1}$)	1208	2114	3308	3414	5759	7444
N applied after 1st cut ($kg\,ha^{-1}$)	60	110	160	60	110	160
Rate of regrowth in 1st month after cutting ($kg\,DM\,ha^{-1}\,day^{-1}$)	108	129	122	91	96	75

hibernated spring generation of tillers is clearly separated from the new summer generation. This phenomenon is often observed after late hay cuts. The other extreme is illustrated by the grazing of pasture varieties of *Lolium perenne*, where elongation is delayed, decapitation is reduced and the new tiller generation overlaps the old one, resulting in a good transition of tiller generations (Table 2). However, even these late flowering and slowly elongating cultivars can present a sharp gap in the generation of new tillers if a heavy cut is taken from a sward to which a large amount of fertilizer N had previously been applied, as illustrated by the data for *L. perenne* cv. Vigor in Table 3, last column. This drastic reduction in tiller number is due partly to the reduction in the number of lateral tillers before cutting, but mostly to the decapitation of 96% of the elongated main tillers during the first cut.

Tiller mortality strongly affects growth rate (Tables 1 and 2). When the majority of tillers is decapitated in a single cut, leaf recovery is retarded for about ten days as the new tiller generation has to be initiated from axillary buds at the bases of the killed tillers [3]. This gap in the generation of new tillers causes a short but distinct depression in growth, as illustrated by the data for the second cut of *L. perenne* cv. Melino and *Phl. pratense* cv. Erecta (Table 1) and *L. perenne* cv. Vigor (Table 2). The gap between tiller generations is, in fact, the cause of 'midsummer depression in grassgrowth', a phenomenon observed in many cutting trials [5, 8]. High rates of regrowth in midsummer are observed only when there is a smooth transition of tiller generations: after defoliation regrowth starts immediately from existing apices, e.g. *L. perenne* cv. Vigor under grazing management (Table 2).

Summer mortality

A gap in the generation of new tillers always results in a depression in growth. In most cases this depression will be short and transitional. However, in some conditions tiller initiation is defective and few new tillers are produced, resulting in patchy regrowth and invasion of adventive species, as discussed below. This 'summer mortality' is particularly common after a heavy spring crop (induced by high N application and favourable growing conditions), when most of the main tillers are elongated and decapitated during cutting and a substantial proportion of the lateral tillers has died (Table 3). Under such conditions, reserves of carbohydrates in the stubble and roots are low, as discussed later. Tedding and wheeling after cutting also increase summer mortality.

Differences in the degree of summer mortality have been observed between species and cultivars. This is clearly illustrated by comparing mortality and rates of growth of swards of *Phl. pratense* and *L. perenne* receiving 400 kg N $ha^{-1} yr^{-1}$ (Table 1). *Lolium multiflorum* has also been studied under management as severe as that applied to swards of *L. perenne* but appears to be unsusceptible to summer mortality at least when the grass is removed immediately after cutting. On the other hand, *L. multiflorum* suffers more than *L. perenne* from the mechanical stresses of field operations (tedding and wheeling) and its winter hardiness is reduced by high N levels, as illustrated by the mortality of tufts after the severe winter of 1978–79 (Table 4). Under the same conditions no tufts of *L. perenne* were killed.

Table 4. Mortality of *Lolium multiflorum* after the winter 1978–79 as affected by cutting frequency and rate of application of fertilizer N (kg $ha^{-1} yr^{-1}$)

	dead + live tufts/m²			% dead tufts		
	300 N	550 N	800 N	300 N	550 N	800 N
4 cuts in 1978	86	72	63	15	38	40
5 cuts in 1978	94	82	69	4	21	44

Exhaustion of nonstructural carbohydrates is one of the main causes of summer mortality. The gap between tiller generations delays leaf expansion for about ten days. New growth is initiated from axillary buds which are dependent on root and stubble reserves of carbohydrate until at least one fully expanded leaf has been formed [6]. Application of fertilizer N also depresses the content of nonstructural carbohydrates [7]. Hence, N not only increases the gap between tiller generations but also enhances the energy stress of the sward after defoliation. The effect is compounded by the smaller amounts of available carbohydrate in the roots and stubble of spring grass as compared to summer grass. This is illustrated by the data in Table 5. In this experiment, the

Table 5. Percent fructose in DM, after hydrolysis of fructosans, in stubble and roots following total defoliation of simulated spring and summer grass, both subsequently grown under the same conditions; *Lolium perenne* cv. Vigor

Days after defoliation	Stubble		Roots	
	spring grass	summer grass	spring grass	summer grass
0	13.4	27.6	3.2	5.4
4	3.6	5.7	1.0	2.3
10	0.9	12.2	1.1	2.0
24	4.8	16.6	2.7	4.5
43	23.3	30.9	5.1	9.0

fructosan content of roots and stubble was followed after defoliation of simulated spring and summer swards. The former had been hibernated and the latter exposed to artificial summer conditions (10–12 °C and 20 h illumination per day). Both treatments were placed outdoors on 25 March and cut on 9 April and again on 22 April. Regrowth was good and immediate in both cases. Nevertheless, the spring grass passed a near zero level of available carbohydrate ten days following defoliation (Table 5). This observation begs the question of what happens under the conditions of more severe physiological stress such as cutting after high N fertilization and good growing conditions.

Invasion of adventive species

Certain conditions of fertilizer application and defoliation management can readily alter the composition of monoculture swards. This is illustrated by data in Table 6. A pure sward of *L. perenne* was treated from 1976 with three rates of fertilizer N (300, 550 and 800 kg $ha^{-1} yr^{-1}$) combined with an 'early' or 'late' cut of the spring growth (mean dates: 7 May and 17 May). The botanical composition was

Table 6. Botanical composition (relative soil covering) after eight years of a sward, originally comprising *Lolium perenne* cv. Vigor, as affected by time of cutting in the spring and the rate of application of fertilizer N. Lp = *Lolium perenne*, Pa = *Poa annua*, Sm = *Stellaria media*, Er = *Elymus repens*. Observations were made in quadrats on 18 August 1983

Time of spring cut	Rate of applied N kg $ha^{-1} yr^{-1}$	Composition of sward %			
		Lp	Pa	Sm	Er
late[a]	300	100	0	0	0
late	550	56	4	1	39
late	800	3	3	4	90
early[b]	800	98	1	0	1

a mean date : 17 May
b mean date : 7 May

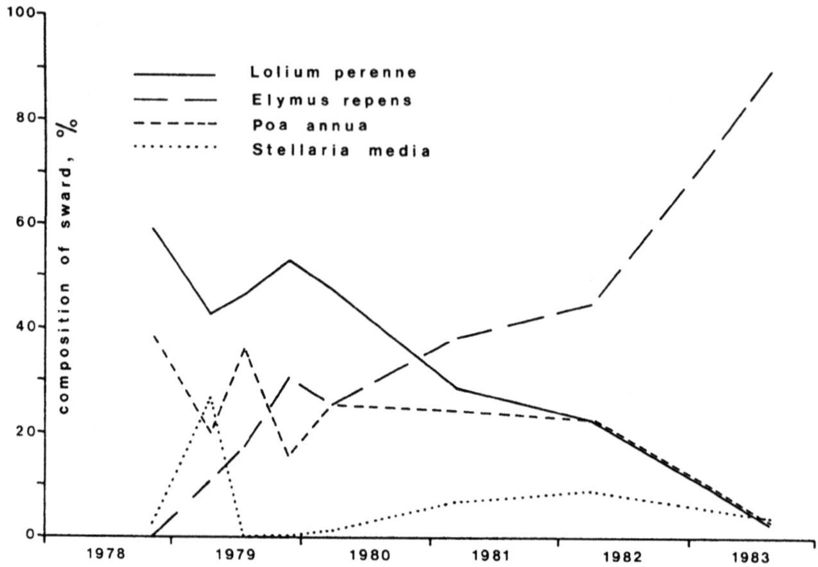

Figure 2. Changes in botanical composition of an originally pure *Lolium perenne* sward (sown 1975, cv. Vigor), under a management of heavy spring cuts (mean date: 17 May) and a high rate of N application (800 kg N ha^{-1} yr^{-1}).

followed on permanent quadrats from 1978, when summer mortality was first observed on some plots. Data in Table 6 indicate that *L. perenne* is readily killed under certain conditions, particularly for late cuts following high rates of N application, and replaced by adventive species. In the first two to four years, bare patches in the swards were invaded by *Poa annua* and *Stellaria media*. In following years, the principle adventive species was *Elymus repens* (syn. *Elytrigia repens*), which became dominant under conditions of repeated summer mortality (Fig. 2). Similar changes in sward composition were observed in an earlier trial: *Phl. pratense* was completely replaced by *E. repens* after 6 years at 500 to 600 kg N ha^{-1} and seven cuts per year. This did not occur at lower N levels.

Conclusions

Cutting a sward with elongated main tillers invariably results in physiological stress arising from disruption in the normal transition of tiller generations. The stress is greatly enhanced by a late spring cut from a sward heavily fertilized with N. These conditions decrease the survival of tillers and thereby the vigour to restore a dense sward after the cut. The sward is thinned and sometimes tufts are actually killed leaving bare patches within the sward. The open space is rapidly invaded by *P. annua*, *S. media* and ultimately *E. repens*. Summer mortality of grass is accentuated by physical factors other than cutting *per se* including wheeling, tedding and covering by cut grass. In order to minimise summer mortality, grass receiving high rates of fertilizer N must be cut before the DM yield reaches 4000 kg ha^{-1} and must be removed from the field as quickly as possible.

References

1. Behaeghe T J (1974) Experiments on the seasonal variation of grass growth. Proceedings of the 12th International Grassland Congress, Moscow 1, 268–281
2. Behaeghe T J (1978) The seasonal discrepancies between potential and actual grass growth. Proceedings of the 7th General Meeting of the European Grassland Federation, Gent 2, 1–20
3. Behaeghe T J (1979) De seizoenvariatie in de grasgroei. Thesis, Rijksuniversiteit, Gent, 272 pp
4. Langer R H M (1972) How grasses grow. London: Edward Arnold
5. Rappe C (1951) Seasonal variations in the rate of pasture regrowth after grazing. Plant and Soil 3, 309–338
6. Sheard R W (1973) Organic reserves and plant regrowth. In Butler G W and Bailey R W, eds. Chemistry and biochemistry of herbage, vol. 2, pp 353–377. London: Academic Press
7. Smith D (1973) The nonstructural carbohydrates. In Butler G W and Bailey R W, eds. Chemistry and biochemistry of herbage, vol. 1, pp 105–155. London: Academic Press
8. Hart M L 't (1948) De seizoenschommelingen van de grasgroei in Nederland. Maandblad Landbouwvoorlichting sdienst 5. 4, 137–148

Sources and transformations of organic nitrogen in intensively managed grassland soils

DC WHITEHEAD
The Animal and Grassland Research Institute, Hurley, Maidenhead, Berkshire, SL6 5LR, UK

Key words: nitrogen, plant residues, roots, faeces, urine, mineralisation, immobilisation, availability

Abstract. Large amounts of nitrogen (200–350 kg N ha^{-1} yr^{-1}) are returned to the soil under intensively managed grassland through the death and decay of unharvested herbage and of roots. With grazed swards, the return of animal excreta may contribute an additional 150–300 kg N ha^{-1} yr^{-1}, but this is distributed unevenly. Earthworms are important in the early stages of decomposition of plant residues and faeces. The balance between the mineralisation and immobilisation of N during decomposition is influenced largely by the C:N ratio. In many grassland soils, the rate of addition of organic matter exceeds the rate of decomposition but, when long-term grassland is subjected to intensive management, this pattern may be reversed. Soils differ considerably in the proportion of the total soil N that is mineralised, and hence becomes available for uptake, during a growing season. Available soil N may be predicted by chemical analysis based on the measurement of NH_4^+-N + NO_3^--N extracted by boiling with 1*M* KCl. In the field, prediction is improved when the results of soil analysis are adjusted by a factor, calculated for the individual site, derived from mean values of temperature and either soil water status or rainfall.

The aim of this paper is to assess, quantitatively so far as possible, some of the sources and transformations of organic N in soil under intensively managed grassland. Particular attention is given to the amounts of N returned to the soil in plant residues and animal excreta, to the balance between immobilisation and mineralisation and to the prediction of the supply of available soil N. For this paper, intensively managed grassland is regarded as that producing a harvestable yield of more than 10 000 kg dry matter (DM) per ha, either from grass swards by the use of fertilizer N at rates of up to 450 kg N ha^{-1} yr^{-1}, or from grass/clover swards relying on symbiotic N fixation.

Additions of N in plant residues and animal excreta

Unharvested leaf and stem material

The amount of unharvested leaf and stem material undergoing death and decay is very variable and it is also difficult to measure. In cut swards, the amounts of stubble left after cutting give some indication of minimum amounts since grass tillers, particularly in intensively managed swards, generally live for less than one year. With a cutting height of 5 cm, the weight of stubble may amount to 3000 kg DM ha^{-1} [1, 9]. Additional unharvested material transferred to the soil results from the

H G van der Meer, J C Ryden and G C Ennik (eds) Nitrogen Fluxes in Intensive Grassland systems.
 ISBN 978-94-010-8454-3

death and decay of leaves during periods of regrowth, and amounts will be influenced by the time interval involved and by weather conditions. With regard to weather, Bowman *et al.* [5] found in Australia that drought was the main factor affecting the rate of senescence of a perennial ryegrass/subterranean clover sward and that, during early summer, senescence often amounted to 2–3% of the standing crop per day. However the decay rate of the dead herbage was related to rainfall, though with a lag period of about 2 weeks, this presumably reflecting the time required for incorporation by the soil fauna and/or the development of appropriate microbial populations. Low temperature and lack of light are also likely to increase the rate of senescence and, in this context, it has been reported from Switzerland that the weight of white clover stolons declined by 30–56% during the winter [32].

A major difficulty in assessing the amount of herbage undergoing death and decay is that senescent herbage is often subject to rapid decomposition or consumption by earthworms and other soil fauna. However it has been estimated that, with both cut and grazed swards under good management, an amount of herbage, almost equal to the amount harvested or consumed, is lost by death or decay during the growing season [19]. With swards cut infrequently, and with grazing intensities lower than the optimum, larger proportions of the total herbage grown undergo death and decay [19, 20]. Additional amounts would be lost by death and decay during the winter.

The concentration of N in the DM of dead herbage material can vary over a wide range from less than 1.0% to more than 3.0%. With grasses, the concentration depends to a large extent on the rate of fertilizer N applied during the previous growth period. With herbage from plots of ryegrass receiving fertilizer N at rates from nil to 450 kg N $ha^{-1} yr^{-1}$, the concentration of N in dead brown herbage collected in August increased with fertilizer rate from 1.08 to 2.48%, while the concentration in green herbage increased from 2.23 to 3.37% (Whitehead, unpublished). Corresponding samples collected in October showed increases of from 1.08 to 2.20% N in dead brown herbage and from 3.05 to 5.50% in living green herbage. The greater difference in October (following greater rainfall than in August) between dead and living herbage at the high rate of fertilizer N suggests that some of the loss of N during leaf senescence was due to leaching by rainfall. In New Zealand, Hunt [13] found that the dead herbage of perennial ryegrass from plots receiving 155 kg N $ha^{-1} yr^{-1}$ contained an average of 1.5% N, while similar material from plots receiving the extremely high rate of 2360 kg N $ha^{-1} yr^{-1}$ contained 3.2% N. With white clover, dead leaf + petiole has been found to contain 2.0–3.5% N while the dead herbage

Table 1. Estimates of the amounts of N returned to the soil in unharvested herbage, roots, faeces and urine

	Grass sward receiving fertilizer N		Grass/clover sward	
	Cut	Grazed	Cut	Grazed
Assumed yield of harvested herbage, kg DM ha^{-1} yr^{-1}	12000	12000	10000	10000
Unharvested herbage				
kg DM ha^{-1} yr^{-1}	9000	10000	7000	9000
% N	2.0	2.0	2.0	2.0
kg N ha^{-1} yr^{-1}	180	200	140	180
Roots				
kg DM ha^{-1} yr^{-1}	6000	6000	6000	6000
% N	1.5	1.5	1.8	1.8
kg N ha^{-1} yr^{-1}	90	90	108	108
Faeces (85% excreted at pasture)				
kg DM ha^{-1} yr^{-1}	–	3060	–	2550
% N	–	2.5	–	2.5
kg N ha^{-1} yr^{-1}	–	77	–	64
Urine (85% excreted at pasture)				
l ha^{-1} yr^{-1}	–	17850	–	14880
g N l^{-1}	–	10	–	9
kg N ha^{-1} yr^{-1}	–	179	–	134

of grass grown with clover but without fertilizer N contained only 0.9–1.3% N (Whitehead, Ryden and Parsons, unpublished).

Estimates of the amounts of N transferred to the soil through the death and decay of herbage from different swards are given in Table 1.

Roots

The amount of N transferred each year from roots to soil organic matter is influenced by the amount of roots produced, their length of life and their concentration of N. A newly-sown grass sward may produce up to about 3000 kg root organic matter per ha in 6 months, increasing to 10 000 kg ha^{-1}, including both living and dead roots, after 3 or 4 years [9].

There is apparently conflicting information on the influence of fertilizer N on root production. In some investigations [e.g. 25] N has been reported to increase root growth and, in other investigations [e.g. 7, 18] a decrease has been reported. A possible explanation is that grass roots respond to increasing amounts of N by increased growth up to a certain amount and then by decreased growth [7], a trend illustrated by the results obtained with ryegrass grown in pots by Toomre [29]. The N supply at which root growth reaches a maximum is probably well below that producing maximum herbage growth.

There is evidence that when swards are defoliated at intervals of 3–4 weeks, the average length of life of grass roots is about 5–6 months [9, 30], although in the absence of defoliation it may be more than 1 year. The death of roots is increased by stress factors such as defoliation [4, 18] and therefore there will be a greater turnover of root material under intensively managed, than under natural or semi-natural grasslands. A major problem in assessing the average length of life of grass roots is the difference between the vascular stele and the cortex. Roots will continue to grow, and to absorb and transport water and nutrients, even though, in the older region of the root, the cortex has disintegrated and decomposed. Even when a root is completely dead, the cortex will decompose much more rapidly than the stele. Taking these factors into account, it is likely that the annual transfer of material from roots to soil organic matter under an intensively managed grass sward will be about 6000 kg DM ha^{-1} (Table 1). The weight of roots under white clover grown alone is considerably less than that under perennial ryegrass, probably due in part to a greater rate of decomposition [36], but root weights under mixed grass/clover swards may be as high as those under grass alone [9].

The concentration of N in grass roots is very much influenced by the supply of fertilizer N. Minimum concentrations may be less than 1.0% in the root organic matter, but concentrations increase with fertilizer N and may approach 2.0% [33], considerably less than the concentrations in herbage at high rates of fertilizer N. The roots of legumes contain considerably more N than those of grasses, even grasses that have received high rates of fertilizer N. Thus white clover roots, including nodules, normally contain between 2.5 and 4.0% N in the organic matter [34, 36]. However the content is reduced by stress factors such as frequent defoliation and drought [36], probably due mainly to the loss of root nodules.

Animal excreta

If herbage amounting to 12 000 kg DM ha^{-1} yr^{-1} is consumed by grazing animals, the amount of faeces excreted would be 3600 kg DM ha^{-1} yr^{-1} assuming 70% digestibility of the herbage. The N content of the faeces of cattle and sheep fed on grass or grass/clover is usually in the range 2–3% of dry weight [33]; with a content of 2.5%, this amount of faeces would contain 90 kg N. For a grazing intensity of 700 cow days ha^{-1} yr^{-1}, 6.4% of the grazed area would have been covered by faeces. This assumes an average of 12 faecal excretions per cow per day, 85% of which occur at pasture and each covering an area of 0.09 m^2 [39]. Similar data have been reported by MacDiarmid and Watkin [14]. Estimates of the rate of N added to the area actually covered by faeces include 850 kg N ha^{-1} [21] and 1330 kg N ha^{-1} [17]. For the assumptions outlined above, the return of N on the area actually covered would be 1200 kg ha^{-1}. Of the faecal N, about 25% is generally in water-soluble forms representing metabolic products of the animal and the microbial population in the gut, about 10–20% is undigested dietary N and the remainder, amounting to about 60%, is present in bacterial cells [16].

For animals in intensive grassland systems, the excretion of N in the urine is greater than that in the faeces. With herbage containing more than 4% N, Barrow and Lambourne [2] found that about 80% of the N excreted by sheep was present in the urine, whereas with herbage containing about 0.8% N the proportion was only 43%. The concentration of N in urine varies widely with factors such as the N content in the diet and the consumption of water, but the data that are available suggest an average concentration of 8–10 g N l^{-1} for both cattle and sheep [33]. There are a number of reports of urination frequency per day, but very few measurements of volumes produced particularly under grazing conditions. However, again assuming a grazing intensity of 700 cow days ha^{-1} yr^{-1}, a urination frequency of 10 times per day [10, 15, 21], with 85% at pasture, an average urination volume of 3 l [3] and a concentration of 10 g N l^{-1}, then the return of N in urine would be equivalent to 180 kg N ha^{-1} yr^{-1} (Table 1). These values for urine production and N concentration clearly represent averages from a wide spectrum, within which actual values will be very much influenced by factors such as size of animal, diet and the intake of water. With high-yielding swards producing herbage containing > 3% N, there are reports that the amount of N returned in urine may be as high as 240 kg N ha^{-1} yr^{-1} [17, 31]. This figure is consistent with the consumption by grazing animals of 12 000 kg DM containing 3.3% N, and a urinary excretion of 60% of the N intake.

Assuming a grazing intensity of 700 cow days ha^{-1} yr^{-1}, with an average of 10 urinations per cow per day, 85% at pasture and each urination covering an area of 0.28 m^2 [33] with no overlapping, 16.7% of the grazed area would receive urine during the year. Hence, if the total return amounted to 180 kg N ha^{-1} yr^{-1}, the area directly covered by urine would receive 1080 kg N ha^{-1}. If the total return was 240 kg N ha^{-1} yr^{-1} and again only 16.7% of the total area was covered, this proportion of the area would receive 1440 kg N ha^{-1}. However, the area influenced by the return of urine is often more than twice the area actually covered [33].

Decomposition of organic residues and mineralisation/immobilisation of nitrogen

The role of earthworms and other invertebrates

Although the decomposition of plant residues and animal excreta is essentially a microbiological process, the soil fauna, particularly earthworms, are important in the early stages of decomposition. The soil fauna partially digest the organic residues that they consume, and also transfer organic residues from the soil surface into the bulk of the soil, where they are more accessible to the soil micro-organisms.

Grassland soils generally contain many more earthworms than soils under cultivation [27], in a comparison at Rothamsted about ten times more [6]. Also grazed plots of grassland have been reported to contain more than cut plots [11]. The intensification of grassland production by the use of fertilizers, and by increasing the numbers of grazing animals, therefore tends to increase earthworm populations by increasing the amounts of plant residues and faeces available as substrate. Several studies have shown that earthworm casts are enriched in N relative to the surface soil and that there is a small increase in the proportion of plant-available N [24, 28].

Results from a study in New Zealand have indicated that, for a permanent pasture, more than 6000 kg DM ha^{-1} yr^{-1} as dead leaf litter was consumed by earthworms and that more than 70% of the N in the litter accumulated in the worm casts [28].

Mineralisation/immobilisation involving plant residues and animal excreta

The processes of mineralisation and immobilisation occur continuously in all soils, and the balance between them is influenced mainly by the C:N ratio of the materials undergoing decomposition. With a ratio of 25:1 to 30:1, there may well be net immobilisation in the first few days or weeks of decomposition after which there will be net mineralisation. With C:N ratios greater than 30:1, net immobilisation is likely to be more prolonged, and with ratios of less than 25:1 mineralisation is likely to occur more rapidly. While the C:N ratio is a major factor, other aspects of composition also influence the rate of decomposition, and hence the mineralisation of N. For example, the relative proportions of lignin and carbohydrate in plant residues were found to be important by Herman *et al.* [12] who reported that the rate of decomposition was inversely related to:

$$\frac{(\text{C:N ratio}) \times (\%\ \text{lignin})}{(\sqrt{\%\ \text{carbohydrate}})}$$

With rates of fertilizer N in the range 350–450 kg N ha^{-1} yr^{-1}, grass roots have C:N ratios of around 25–30:1 [33] and dead grass herbage ratios of <25:1. Clover roots and herbage usually have ratios in the range 12:1 to 20:1. However, dead herbage and roots of grass grown with little or no fertilizer N, including that grown in association with clover, generally have C:N ratios of between 40:1 and 50:1. In grassland receiving moderately high rates of fertilizer N, therefore, plant residues are unlikely to immobilise through microbial activity any appreciable amount of N during their decomposition. Nevertheless when fertilized residues contribute to soil organic matter, there is some conversion of fertilizer N to organic forms. The return of faeces and urine derived from fertilized herbage has a similar effect, though with urine the effect is short-lived since mineralisation is rapid. In swards receiving little or no fertilizer N, there may be some microbial immobilisation of soil inorganic N during the decomposition of grass, though not clover, residues. Immobilisation by unfertilized grass roots has been shown to occur by Power [22].

With many grassland soils, particularly those previously subject to periods of cultivation, the rate of addition of organic matter is greater than the rate of decomposition, and consequently the organic matter content of the soil tends to increase, and this increase may continue for many years if the sward remains unploughed. Because the C:N ratio of humified soil organic matter varies only within narrow limits around 10.1, the N content of the soil also tends to increase with time. Ultimately under uniform management, an equilibrium is reached when additions of organic matter and N are balanced by mineralisation and losses. However changes in management will influence the equilibrium: for example fertilizer N and lime often tend to increase the rate of decomposition of the soil organic matter. Thus while large applications of fertilizer N will increase the amount of N contributed to the soil organic matter through unharvested plant material and through animal excreta, the rate of decomposition of organic material may increase and exceed the rate of addition. Recent results from New Zealand have suggested that, with the intensification of use of long-established grassland, there may be a decline in the soil contents of organic matter and N [8]. It is likely that there will also be some re-distribution of N within the soil profile following the intensification of use of long-established low-production grassland. Factors such as a moderate increase of fertilizer use, liming and/or drainage may well encourage a greater depth of rooting and an increase in the population of earthworms, so that more organic material will be transferred to the soil at depths below 15 cm. At the same time, the rate of decomposition of organic matter accumulated at the soil surface will increase.

Availability of soil nitrogen

Differences in availability between soils

The soil N taken up by grass over the course of a growing season includes (i) inorganic N ($NH_4^+ + NO_3^-$) present in the soil at the beginning of the season and (ii) N released from the potentially mineralisable organic N during the season. Because established grassland has a well-developed and extensive root system, the amount of N taken up from plots receiving no fertilizer N provides a good indication of the available soil N supply. Using this method of assessment, and including results from a large number of experiments in England and Wales, Richards and Hobson [23] concluded that soil N supply varied from 8 to 350 kg N $ha^{-1} yr^{-1}$. As well as these large differences in absolute amounts, there are also large differences between soils in the proportions of the total soil N made available over a growing season. This is shown by data from an experiment in which 21 different soils were compared [37]. The soils were all obtained from field plots that had received no fertilizer N in the year before sampling, and perennial ryegrass was grown in pots under controlled conditions of light, temperature and, so far as possible, water supply. The percentage of the soil N in the pots that was taken up by the grass over six harvests, and including the N in roots and stubble, varied between the soils from 1.5 to 4.0%. Preliminary results of a subsequent investigation indicate that the amount of N in the 'light fraction', comprising partially decomposed organic matter, is an important factor influencing soil N supply.

Prediction of soil N supply

Under grass swards in humid temperate regions, most of the available soil N taken up during the growing season comes from the potentially mineralisable organic N rather than the inorganic N present at the beginning of the season, but the proportion will vary with soil type, weather and sward management. Therefore, to be generally applicable, any method for predicting soil N supply must include both the inorganic and potentially mineralisable fractions, measured either separately or together. The inorganic N, although usually present in grassland soils in small amounts, is easy to measure and is completely available for uptake by plants. The potentially mineralisable N is more difficult to measure and, in the field, is only partially available because factors such as temperature and water status are usually less than optimum for mineralisation. The organic N in soils consists of several fractions, including fresh residues from plants and animal excreta, microbial biomass and residues, and humified organic matter. These fractions, although not entirely distinct from one another, differ in their susceptibility to mineralisation, with humified organic matter being the least susceptible.

The most accurate way to estimate the potentially mineralisable N is to measure the NH_4^+-N released during a period of incubation, but methods involving incubation are too time-consuming for routine use. Various more rapid chemical extraction methods have therefore been examined and the results compared with the amounts of N taken up by ryegrass [35]. This comparison involved 21 soils and the ryegrass was grown in pots under controlled environment conditions. Of the various methods, the best in terms of predicting the available N supply, was to measure the amount of NO_3^--N + NH_4^+-N extracted by boiling a sample of soil with 1*M* KCl [35]. This procedure completely extracts the inorganic N, and in addition releases some NH_4^+-N from the organic matter providing an estimate of the potentially mineralisable N.

It has been known for many years that temperature and water status influence the rate of mineralisation of the potentially mineralisable soil organic N. However these factors had not been taken into account in the prediction of soil N supply until Stanford *et al.* [26] suggested the possibility of calculating an adjustment factor for any particular site that could be applied to the results of soil analysis. To calculate the factor, the growing season was divided into periods of 4 weeks and mean values obtained for soil temperature and water status, expressed as a percentage of available water capacity. The adjustment factor was then calculated on the basis of relationships (i) between mineralisation and temperature and (ii) between mineralisation and water status, both derived from incubation studies. However Stanford *et al.* gave no indication whether the factor improved the prediction of soil N supply in their investigation, which involved sugar beet grown at a range of sites in Maryland, USA, but calculations from information given in their paper indicate that it did not achieve any improvement. Nevertheless, a similar adjustment applied in an investigation involving ryegrass grown at a number of sites in the UK did substantially improve the prediction of soil N supply using chemical extraction methods [38].

Data on air temperature and rainfall are generally more readily available than data on soil temperature and soil water status, and can be used to derive a simpler adjustment factor (F).

$$F = \frac{t \times r}{k}$$

where t = mean air temperature in °C, April–September inclusive, r = mean total rainfall in mm, April–September inclusive, and k is a constant which provides for the conversion of soil analysis results, expressed in mg N kg^{-1} soil, to soil N supply expressed in kg N ha^{-1} yr^{-1}. On the basis of the limited amount of data available to date, k has been given a provisional value of 3000 (the optimum value for k will be determined on the basis of future results).

In order to obtain the predicted value for soil N supply (kg N ha^{-1}), the result of soil anaysis (mg NO_3^--N + NH_4^+-N kg^{-1} less soil, 0–15 cm, extracted by boiling with 1*M* KCl) is multiplied by the value for F for the particular site. However, the extent to which prediction is improved by the use of the adjustment factor, F, or a modification of it requires further investigation.

Conclusions

In intensively managed grassland swards, large amounts of N are transferred to the soil through the death and decay of plant residues, both shoots and roots, and with grazed swards through the return of animal excreta. The rate of decomposition of these materials and hence the availability for plant uptake of their N content varies widely, from urine in which the N is rapidly available, to grass roots grown without fertilizer N (though possibly in association with clover) in which the N is slowly available. Soils differ considerably in the proportion of the soil organic N mineralised annually, even under uniform environmental conditions, and the reasons for such differences require elucidation. More effective ways of predicting the supply of available N from the soil over the course of a growing season would improve the efficiency of use of fertilizer N.

References

1. Baker H K and Garwood E A (1959) Studies on the root development of herbage plants. IV. Seasonal changes in the root and stubble weight of various leys. Journal of the British Grassland Society 14, 94–104
2. Barrow N J and Lambourne L J (1962) Partition of excreted nitrogen, sulphur and phosphorus between the faeces and urine of sheep being fed pasture. Australian Journal of Agricultural Research 13, 461–471
3. Blagden P (1969) The potassium cycle in grassland. Ph.D. thesis, University of Nottingham
4. Boote K J (1976) Root:shoot relationships. Proceedings Soil and Crop Science Society of Florida 36, 15–23
5. Bowman P J, White D H, Cayley J W D and Bird P R (1982) Predicting rates of pasture growth, senescence and decomposition. Proceedings of the Australian Society of Animal Production 14, 36–37
6. Edwards C A and Lofty J R (1972) Biology of Earthworms. London: Chapman and Hall, pp 283
7. Ennik G C, Gillet M and Sibma L (1980) Effect of high nitrogen supply on sward deterioration and root mass. In Prins W H and Arnold G H, eds. The role of nitrogen in intensive grassland production, pp 67–76. Wageningen: PUDOC
8. Field T R O and Ball P R (1981) Nitrogen balance in an intensively utilised dairy farm system. Proceedings of the New Zealand Grassland Association 43, 64–69

9. Garwood E A (1967) Studies on the roots of grasses. Annual Report 1966, pp 72–79. Hurley, UK: The Grassland Research Institute
10. Hancock J (1953) Grazing behaviour of cattle. Animal Breeding Abstracts 21, 1–13
11. Heath G W (1962) The influence of ley management on earthworm populations. Journal of the British Grassland Society 17, 237–244
12. Herman W A, McGill W B and Dormaar J F (1977) Effects of initial chemical composition on decomposition of roots of three grass species. Canadian Journal of Soil Science 57, 205–215
13. Hunt W F (1983) Nitrogen cycling through senescent leaves and litter in swards of Ruani and Nui ryegrass with high and low nitrogen inputs. New Zealand Journal of Agricultural Research 26, 461–471
14. MacDiarmid B N and Watkin B R (1972) The cattle dung patch. III. Distribution and rate of decay of dung patches and their influence on grazing behaviour. Journal of the British Grassland Society 27, 48–54
15. MacLusky D S (1960) Some estimates of the areas of pasture fouled by the excreta of dairy cows. Journal of the British Grassland Society 15, 181–188
16. Mason V C, Kessank P, Ononiwu J C and Narang M P (1981) Factors influencing faecal nitrogen excretion in sheep. 2. Carbohydrate fermentation in the caecum and large intestine. Zeitschrift für Tierphysiologie, Theirer nährung and Futtermittelkunde 45, 174–184
17. O'Connor K F (1974) Nitrogen in agrobiosystems and its environmental significance. New Zealand Agricultural Science 8, 137–148
18. Oswalt D L, Bertrand A R and Teel M R (1959) Influence of nitrogen fertilization and clipping on grass roots. Proceedings Soil Science Society of America 23, 228–230
19. Parsons A J (1986) The effects of season and management on the growth of temperate grass swards. In Jones M B and Lazenby A, eds. The grass crop – the physiological basis of production. London, New York: Chapman and Hall (in press)
20. Parsons A J, Leafe E L, Collett B, Penning P D and Lewis J (1983) The physiology of grass production under grazing. II. Photosynthesis, crop growth and animal intake of continuously grazed swards. Journal of Applied Ecology 20, 127–139
21. Petersen R G, Woodhouse W W and Lucas H L (1956) The distribution of excreta by freely grazing cattle and its effect on pasture fertility. 2. Effect of returned excreta on the residual concentrations of some elements. Agronomy Journal 48, 444–449
22. Power J E (1968) Mineralization of nitrogen in grass roots. Proceedings Soil Science Society of America 32, 673–674
23. Richards I R and Hobson R D (1977) The relationship between nitrogen supply and the nitrogen yield of cut grass swards. Journal of Agricultural Science, Cambridge, 88, 289–292
24. Satchell J E (1967) Lumbricidae. In Burges A and Raw F, eds. Soil biology, pp 259–322. London: Academic Press
25. Schuurman J J and Knot L (1974) The effect of nitrogen on the root and shoot development of *Lolium multiforum* var. *Westerwoldicum*. Netherlands Journal of Agricultural Science 22, 82–88
26. Stanford G, Carter J N, Westerman D T and Meisinger J J (1977) Residual nitrate and mineralisable soil nitrogen in relation to nitrogen uptake by irrigated sugarbeets. Agronomy Journal 69, 303–308
27. Steen E (1983) Soil animals in relation to agricultural practices and soil productivity. Swedish Journal of Agricultural Research 13, 157–165
28. Syers J K, Sharpley A N and Keeney D R (1979) Cycling of nitrogen by surface-casting earthworms in a pasture ecosystem. Soil Biology and Biochemistry 11, 181–185
29. Toomre R I (1966) Effect of high rates of nitrogenous fertilizers on the yield and chemical composition of grasses. Proceedings 10th International Grassland Congress, Helsinki, 1966, 227–230
30. Troughton A (1981) Length of life of grass roots. Grass and Forage Science 36, 117–120
31. Van der Meer H G (1983) Effective use of nitrogen on grassland farms. In Corrall A J, ed. Proceedings 9th General Meeting of the European Grassland Federation, Reading, 1982, 61–68

32. Vez A (1961) (Study of the development of white clover in relation to fluctuations in certain organic substances). Berichte der Schweizerischen Botanischen Gesellschaft 71, 118–172
33. Whitehead D C (1970) The role of nitrogen in grassland productivity. Bulletin 48, Commonwealth Bureau of Pastures and Field Crops, 202 pp. Farnham Royal, U.K.: Commonwealth Agricultural Bureaux
34. Whitehead D C (1970) Carbon, nitrogen, phosphorus and sulphur in herbage plant roots. Journal of the British Grassland Society 25, 236–241
35. Whitehead D C (1981) An improved chemical extraction method for predicting the supply of available soil nitrogen. Journal of the Science of Food and Agriculture 32, 359–369
36. Whitehead D C (1983) The influence of frequent defoliation and of drought on nitrogen and sulphur in the roots of perennial ryegrass and white clover. Annals of Botany 52, 931–934
37. Whitehead D C (1984) Interactions between soil and fertilizer in the supply of nitrogen to ryegrass grown on 21 soils. Journal of the Science of Food and Agriculture 35, 1067–1075
38. Whitehead D C, Barnes R J and Morrison J (1981) An investigation of analytical procedures for predicting soil nitrogen supply to grass in the field. Journal of the Science of Food and Agriculture 32, 211–218
39. Wilkinson S R and Lowrey R W (1974) Cycling of mineral nutrients in pasture ecosystems. In Butler G W and Bailey R W, eds. Chemistry and biochemistry of herbage, Volume 2, pp 247–315. London: Academic Press

Gaseous losses of nitrogen from grassland

JC RYDEN
The Animal and Grassland Research Institute, Hurley, Maidenhead, Berkshire SL6 5LR, UK

Key words: field methods, nitrous oxide, denitrification, ammonia volatilisation, grazed swards, nitrogen balance

Abstract. The major processes involved in gaseous losses of nitrogen (N) from grassland are denitrification and ammonia (NH_3) volatilisation. The most widely used methods to measure such losses are based on soil covers and micrometeorological techniques. The former are used primarily in measuring nitrous oxide and denitrification loss in conjunction with ^{15}N or the acetylene-inhibition technique. In studies of NH_3 loss, micrometeorological techniques are generally preferred, the mass balance method being the best suited to most grazing experiments. In systems dependent on NH_4NO_3 fertilizer, denitrification is the major pathway of direct loss from the N applied with annual losses of between 5 and 10% for cut swards. Higher rates of loss are observed in grazed swards, and are to be expected following the application of slurry. Losses of NH_3 from urea fertilizer range from 5 to 20%. Depending on weather conditions, a 4 to 66% loss of N has been observed for urine- and dung-affected areas of pasture whilst losses of 20 to 120 kg N ha^{-1} yr^{-1} have been reported for whole grazed swards. Large losses of NH_3 have also been inferred from measurements following the application of slurry. Direct denitrification loss from NH_4NO_3 fertilizer is reduced substantially when ammoniacal fertilizers are used. Reduction of the NH_3 loss associated with excreta is to be expected with irrigation, amendment of cation exchange capacity in the surface soil and by injection of slurry. Further field studies are required to establish the principal factors affecting gaseous losses of N and to identify appropriate management through which they may be reduced.

The production of gaseous nitrogen (N) compounds is potentially one of the most important pathways of loss affecting the ultimate fate of N in grassland agriculture. The major processes involved are denitrification and ammonia (NH_3) volatilisation. However, smaller but possibly significant losses may arise through nitrous oxide (N_2O) production during nitrification and through emission of volatile amines. Gaseous losses of N may occur directly from fertilizer, from the N voided in excreta and from plant residues during decomposition. In areas where ammonium nitrate is the major fertilizer in use, denitrification is the only process likely to result in significant direct loss of N. Ammonia volatilisation is expected only from urea and possibly ammonium nitrate-lime (calcium ammonium nitrate). In contrast, both NH_3 volatilisation and denitrification are important processes determining the fate of N returned in excreta [3].

This paper outlines methods suitable for field studies of gaseous losses of N. Results from studies in which these methods have been used are summarized and possible methods to reduce gaseous losses are discussed.

H G van der Meer, J C Ryden and G C Ennik (eds) Nitrogen Fluxes in Intensive Grassland systems.
 ISBN 978-94-010-8454-3

Methods for measuring gaseous losses of nitrogen

Methods suitable for field studies of gaseous losses of N fall into three basic groupings:

(i) Those involving diffusion theory to calculate gas transport through the soil from measurements of gas concentrations in the soil atmosphere [e.g., ref. 31].

(ii) Those in which the flux is calculated from changes in gas concentration within covers or enclosures placed over the soil surface [e.g., refs. 38, 22].

(iii) Those in which the flux is determined from micrometeorological measurements and gas concentrations above the soil surface [e.g., refs. 12, 14, 26].

Due to the uncertainties associated with methods under group (i) [29], soil cover and micrometeorological techniques are currently the most widely adopted methods in studies of gaseous losses of N. These techniques have recently been discussed in reviews by Ryden and Rolston [41] and Denmead [11] from which the following conclusions may be drawn:

(i) Micrometeorological techniques are the preferred method in any field study in that they have no effect on the gas flux to be measured. Furthermore, their use is the only possible approach to studies of NH_3 loss from whole grazed swards in which the gas is produced from a heterogeneous distribution of urine- and dung-affected areas. In the case of nitrogen oxides, their concentration in the atmosphere close to the ground surface is usually no more than a few percent above their ambient levels, forcing the adoption of covers or enclosures. With careful attention to design and operating conditions [10, 22], the latter are reliable and sensitive for studies of nitrogen oxide loss and provide a basis for techniques to measure total denitrification in the field. Covers or enclosures are generally considered to be unsuited to studies of NH_3 loss due to the reactivity of this gas with water and the dependence of such loss on prevailing meteorological conditions. However, suitable designs for small plot work have been proposed [27, 37, 46] in which the air speed through the enclosure can be controlled to within the range of wind speeds outside the cover.

(ii) ^{15}N provides the basis for the only direct method by which denitrification can be measured although its use is restricted by the cost of materials and analytical equipment. In most applications [e.g., refs. 30, 31] it has been possible to measure only the N loss occurring from the fertilizer applied. However, recent developments in the analysis of data [42] now permit non-destructive estimation of the pool of nitrate undergoing denitrification and thereby the determination of the total loss of N_2. Although ^{15}N techniques can be used to study denitrification following the application of fertilizer, they are unlikely to be practical in studies of denitrification in soils affected by dung, urine or slurry.

(iii) The acetylene (C_2H_2)-inhibition technique [39] is the most widely applicable method for field studies of denitrification. Although the technique is conceptually simple, various potential problems have been identified. Laboratory studies have shown that (a) C_2H_2 also inhibits nitrification, (b) inhibition of N_2O reduction may become ineffective after long term exposure of soil to C_2H_2 and (c) C_2H_2 may be oxidized leading directly or indirectly to enhanced rates of denitrifaction. However in at least two field applications of the technique, these problems have been shown to be unimportant in repeated short-term measurements at intervals of 1 to 3 days [36]. The technique is sensitive and inexpensive and is particularly suited to studies in grassland where denitrification may arise from nitrate derived from widely different sources.

(iv) Most micrometeorological techniques can be applied only above extensive and uniform land surfaces and are inappropriate to studies of NH_3 loss from plots of the size normally used in grazing experiments. The recently proposed mass balance method [11, 14] overcomes this limitation, the flux being calculated from the difference in the amount of the gas carried in the air flow across the windward and leeward edges of an experimental area. This method eliminates many of the difficulties including stability correction associated with other micrometeorological methods and involves a few, relatively simple measurements. It is ideally suited to the determination of NH_3 loss from whole grazed swards [40].

All the methods outlined above rely on accurate determination of gas concentrations. Analytical procedures for nitrogen oxides and measurements of denitrification usually involve gas chromatography or infra-red gas analysis whilst mass spectrometry is required in denitrification studies using ^{15}N [41]. Ammonia is usually determined by drawing air through dilute acid followed by colorimetric analysis.

Recent developments in methods outlined above permit studies of the factors affecting gaseous losses of N as they occur in the field. Furthermore, in conjunction with other techniques, they provide for the first time an opportunity to determine complete balances for N in representative grassland systems including those involving the grazing ruminant.

Measurements of gaseous losses of nitrogen from grassland

Oxides of nitrogen and denitrification

The measurement of N_2O flux has been the objective in most studies of gaseous N loss. Of those reported, few have direct relevance to intensively managed grassland in temperate regions. Webster and Dowdell [47] reported a consistently greater N_2O loss from a clay soil as opposed to a silt-loam, each sown to ryegrass and receiving 400 kg NO_3^--N ha^{-1} yr^{-1}. During summer, peak fluxes were as high as 0.250 kg N ha^{-1} day^{-1} particularly when application of N coincided with rainfall or irrigation. Similar patterns of N_2O loss have been observed for a clay-loam in south east England sown to ryegrass receiving 250 kg N ha^{-1} yr^{-1} (Fig. 1). Although individual measurements indicated that daily rates of N_2O loss occasionally exceeded 0.400 kg N ha^{-1}, mean rates over intervals of one week never exceeded 0.212 kg N ha^{-1} day^{-1} [32]. Significantly, this soil acted as a sink for atmospheric N_2O when the soil water content exceeded 20% and NO_3^- content fell below 1–2 μg N per g dry soil (Fig. 1). These and other measurements [e.g., refs. 8, 9] suggest that loss of plant available N through N_2O emission from fertilized grassland is low (generally <2% of the input of N) and of little agricultural significance. Measurements of the emission of other nitrogen oxides suggest similarly low losses of N. In Australia, Galbally and Roy [16] observed rates of nitric oxide loss of 0.001 and 0.002 kg N ha^{-1} day^{-1} for ungrazed and grazed pasture, respectively.

The losses of N_2O discussed above followed application of fertilizer N to cut swards and can be attributed largely to denitrification [33, 47]. The observation that N_2O can be produced during nitrification in soil [6] suggests that small but significant loss of N may arise from this process below urine-affected areas of grazed swards. Production of N_2O during nitrification is enhanced by the presence of high concentrations of NH_4^+ and by above neutral pH [5, 6], precisely the conditions prevailing immediately following the deposition of urine. The importance of such loss from urine-affected soil awaits investigation.

Recently, denitrification losses from soils under cut ryegrass receiving 250 to 500 kg N ha^{-1} yr^{-1} have been measured using the C_2H_2-inhibition technique [3, 33]. At most times during the year denitrification proceeded at a rate of less than 0.010 kg N ha^{-1} day^{-1}. Higher daily rates (0.1 to 1.5 kg N ha^{-1}) were observed only when NH_4NO_3 fertilizer was applied to wet soil in spring or summer, and could be attributed to the combined effects of high soil NO_3^- contents and restricted aeration arising from increased soil water content. Following the first application of fertilizer (Fig. 1) only low rates of denitrification were observed as the soil dried to a water content <20%. After the second application, rates were higher and followed the pattern of change in soil water content. However, after about 14 June, 1981, denitrification was restricted by the amount of NO_3^- available for reduction. The soil NO_3^- content had decreased largely as a result of plant uptake. Over the year

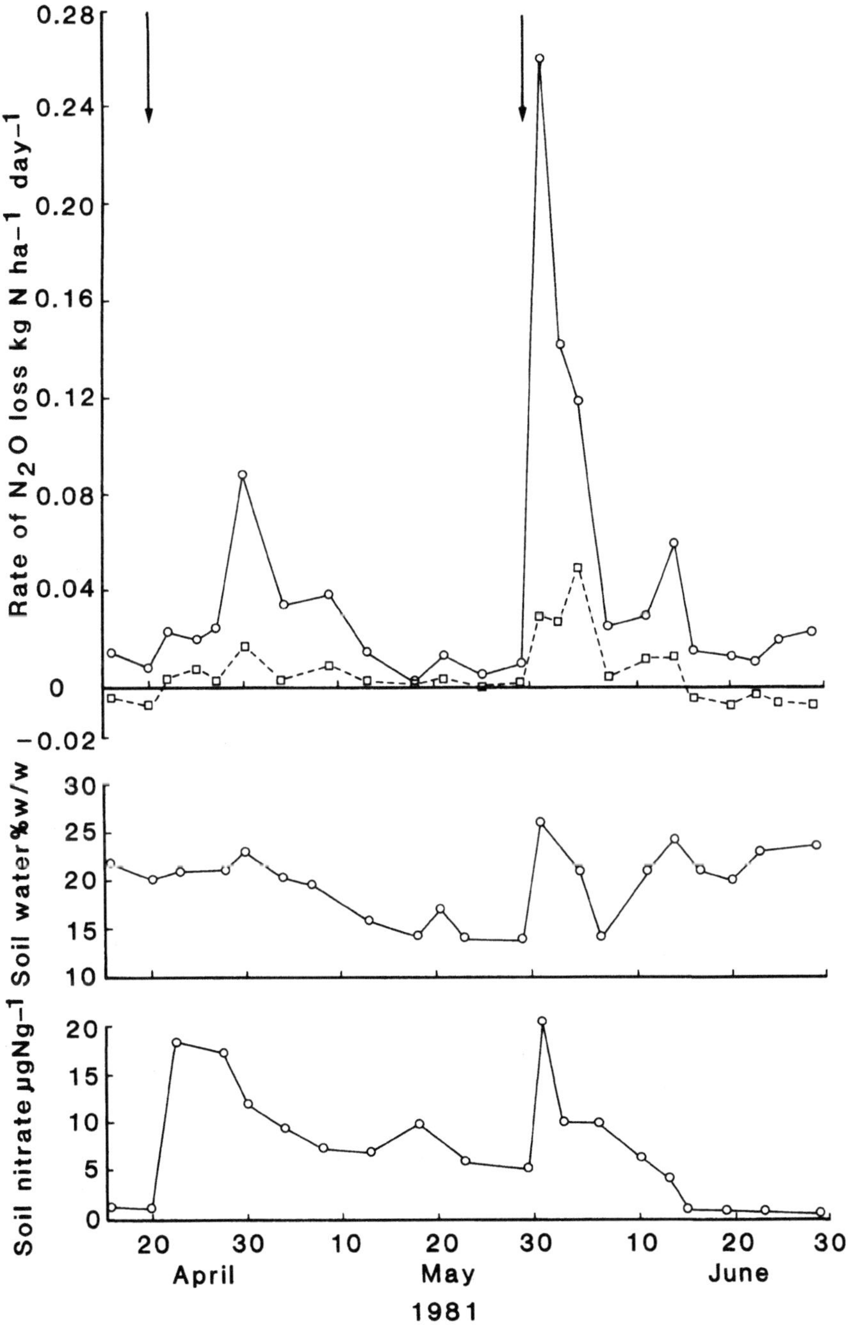

Figure 1. Rates of nitrous oxide loss from a cut ryegrass sward on a clay loam with (○——○) and without (□----□) acetylene treatment and the corresponding soil NO_3^- and water contents (0–20 cm) following additions of NH_4NO_3 fertilizer (62.5 kg N ha^{-1}) on the dates indicated by the arrows. The rate of nitrous oxide loss from acetylene-treated soil is equivalent to the rate of denitrification.

as a whole, denitrification was frequently limited by the soil nitrate content [33].

Limiting values on conditions conducive to higher rates of denitrification (>0.15 kg N ha^{-1} day^{-1}) can be assessed from data such as those in Fig. 1. These are: NO_3^- content >5 μg N per g dry soil (10 to 15 kg N ha^{-1} 20 cm), soil temperature >5°C and an air-filled porosity <15 to 40%. The fact that denitrification was observed at relatively high air-filled porosities reflects the importance of anaerobic microsites to the process as it occurs in the field [19, 44]. The annual denitrification loss from cut ryegrass swards based on individual measurements of the rate of loss is estimated to be between 5 and 10% of the N applied [3, 33].

Larger denitrification losses are expected from grazed swards, particularly towards the end of the grazing season when soil, enriched in inorganic N from excreta [3], returns to field capacity. Even during the summer, mean rates of loss from an irrigated grazed sward were in the range 0.05 to 0.35 kg N ha^{-1} day^{-1}, 1.7 to 16 times greater than those from a comparable cut sward [35]. The higher rates of loss from the grazed sward could be attributed to the higher soil NO_3^- contents in the grazed sward and higher soil water contents in urine- and dung-affected areas, points illustrated by a more recent study in an unirrigated grazed sward (Fig. 2). This indicated that rates of denitrification increased as the soil approached field capacity at the end of the grazing season in October but then decreased in November with decreasing soil temperature. Relative to the cut sward, rates of denitrification below the grazed sward were consistently higher, with rates of loss up to 0.396 kg N ha^{-1} day^{-1} in areas affected by excreta (Fig. 2). The decrease in denitrification rate in November in response to falling soil temperatures indicates that high rates of denitrification may persist for only short periods each year before leaching becomes the dominant process removing excess nitrate from the soil profile [17].

The data currently available suggest that the potential for denitrification loss from grassland soils is high. This reflects the larger amounts of readily oxidisable carbon in grassland soils. Rolston and Broadbent [30] reported that losses from a soil under a cut sward were 2 to 46 times greater than those from the same but fallowed soil. This high potential for loss is realised only in the presence of high NO_3^- concentrations, associated with the application of fertilizer containing NO_3^- or with soils under grazed swards, and is dependent on the development of soil aeration and temperature regimes conducive to denitrification. The rapid uptake of NO_3^- by a growing sward is a major factor limiting loss.

Ammonia volatilisation

Ammonia may be lost directly from fertilizer N but more importantly from excreta returned during grazing or the application of slurry and

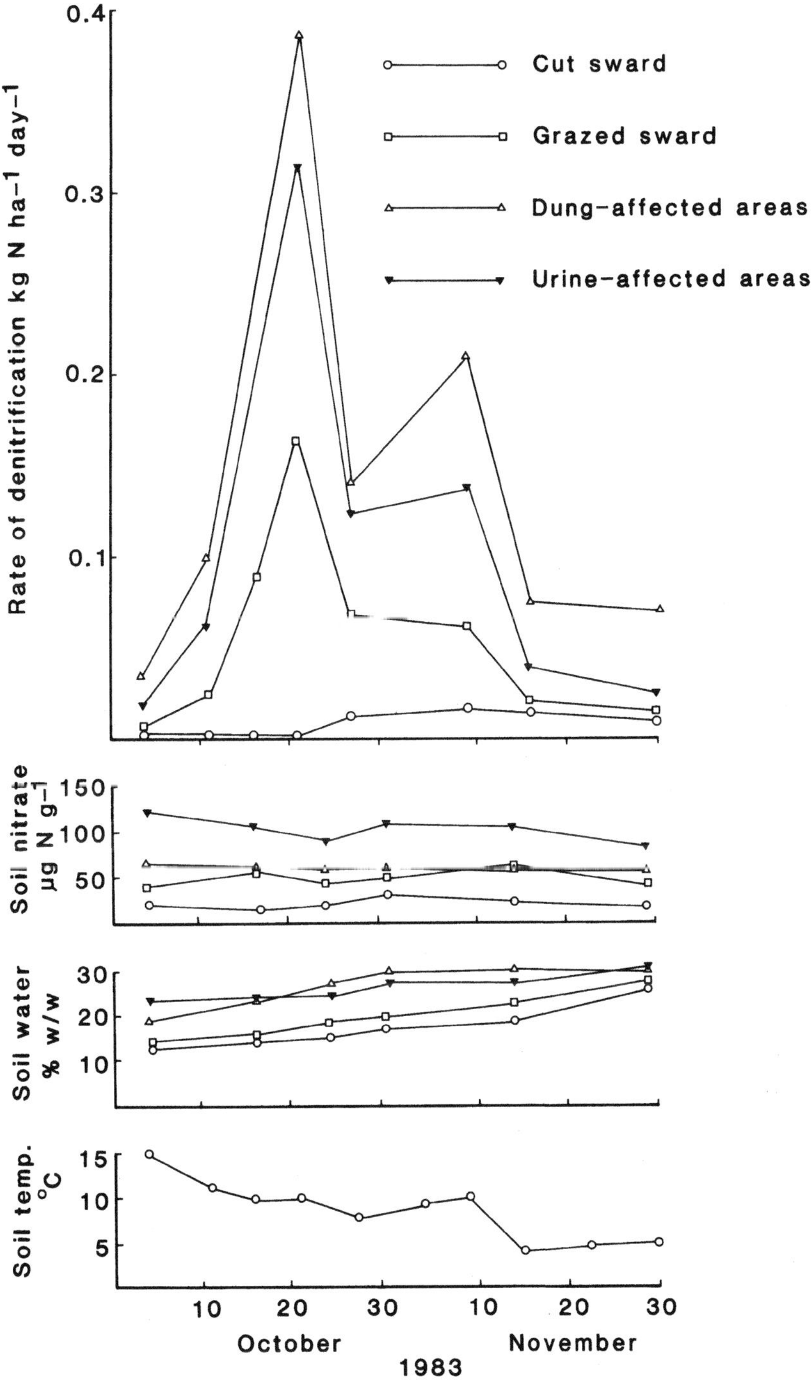

Figure 2. Rates of denitrification loss, soil NO_3^- and water content (0–20 cm) and soil temperature (10 cm) for cut and grazed ryegrass swards including urine- and dung-affected areas of the grazed sward on a freely drained loam approaching field capacity at the end of the season during which 420 kg N ha^{-1} had been applied as NH_4NO_3 to each sward.

other livestock manures. Of the commercially important straight N fertilizers, only the use of urea is likely to result in significant NH_3 loss. Laboratory and field studies [e.g., refs. 18, 20, 21] suggest a 5 to 20% loss of N as NH_3 from urea. The extent of loss is affected by (a) the amount of N applied, (b) soil pH, cation exchange capacity and water content and (c) meteorological conditions affecting evapotranspiration. Smaller but significant loss of NH_3 may also occur from ammonium nitrate-lime (calcium ammonium nitrate), a straight N fertilizer used widely in Germany, The Netherlands and Ireland. Such loss arises from the formation and decomposition of ammonium carbonate and bicarbonate following application of this fertilizer [15]. The importance of such loss awaits investigation.

Urine-N is lost rapidly as NH_3 following hydrolysis of the urea it contains, a process that is largely complete within 24 h of deposition [3]. In studies of the fate of N in urine-treated areas of ryegrass-clover pasture in New Zealand, 66% of the N applied (30 or 60 g N m^{-2}) was lost during warm dry conditions [1, 2]. Losses were lower (6 to 16%) during cooler or wetter conditions. In south east Queensland, Vallis *et al.* [46] reported 14 to 28% loss of the N applied in urine (42.5 g N m^{-2}) depending on the season. In contrast to the large losses of NH_3 from urine, MacDiarmid and Watkin [28] reported a loss of only 4.7% of the N in dung, reflecting a slower hydrolysis of the components containing N most of which are of microbial origin. However, this lower loss of N may reflect the design of the enclosures used which precluded air movement across the ground surface.

There are few data for NH_3 loss from whole grazed fields. Denmead *et al.* [13] observed losses between 0.2 and 0.7 kg N ha^{-1} day^{-1} from lucerne grazed by sheep (50 ewes ha^{-1}) in Australia. They estimated that the annual loss was 100 kg N ha^{-1}. Harper *et al.* [21] reported rates of NH_3 loss of between 0 and 7.5 kg N ha^{-1} day^{-1} from pasture grazed continuously by 4 steers ha^{-1} in south east Queensland. However, many of the highest rates of loss were observed within a few days of the application of urea fertilizer. On some occasions the sward absorbed NH_3 from the atmosphere at rates up to 0.3 kg N ha^{-1} day^{-1}. They estimated that NH_3 volatilisation from urine and dung resulted in an annual loss equivalent to 5.3% of the fertilizer applied (374 kg N ha^{-1}) with a total annual NH_3 loss of 24% of the N applied.

These rates of NH_3 loss from grazed swards contrast with those of 15 to 57 kg ha^{-1} day^{-1} observed above a cattle feedlot in Colorado [23]. Significantly, $<1\%$ of the N loss could be attributed to volatile amines. Although this study has no direct relevance to the European situation, it is illustrative of the large losses of NH_3 that probably occur when cattle are housed during the winter months.

Recent work in the UK (Fig. 3, refs. 3, 35, 40) has shown that NH_3 loss frequently exceeds 1 kg N ha^{-1} day^{-1} during and following grazing of an intensively utilized rotationally grazed ryegrass sward receiving 420 kg N ha^{-1} yr^{-1}. Losses were frequently highest during and immediately after grazing, and decreased during periods of rain and low evapotranspiration [3, 35]. Significantly, NH_3 continued to be lost at rates between 0.2 and 0.5 kg N ha^{-1} day^{-1} up to 24 days after grazing. However, this pattern of loss was observed only during frequent wetting and drying cycles. For example, following the grazing interval in August, 1982 (Fig. 3), the rate of NH_3 loss decreased to rates of approximately 0.25 kg N ha^{-1} day^{-1}. This period of relatively low loss was associated with a dry soil surface, a low mean wind speed and low rates of evapotranspiration. The rate of NH_3 loss then increased with increasing rates of evapotranspiration and passed through a maximum between 12 and 17 August when high rates of evapotranspiration were associated with frequent showers resulting in wetting and drying of the soil surface. Although the NH_4^+ content of the surface soil had decreased to about 10 μg N per g dry soil by this time, the loss of NH_3 continued at about 0.25 kg N ha^{-1} day^{-1} until the beginning of the next grazing period on 29 August.

The relationships in Fig. 3 reflect the many factors that influence NH_3 loss. Although, during August, maximum loss occurred when heavy showers and high wind speeds led to repeated wetting and drying of the soil surface, NH_3 loss often decreased during periods of more persistent rain in May, June and October [3, 35]. The average rates of loss during May/June and August were 0.75 and 1.13 kg N ha^{-1} day^{-1}, respectively, which indicate total losses in the range 20 to 30 kg N ha^{-1} for each of the 28-day grazing rotations. These are substantial when set against the amount of fertilizer applied (60 kg N ha^{-1}) prior to each regrowth interval. A full understanding of the loss of NH_3 awaits more critical evaluation of the soil and environmental factors affecting the process of volatilisation, especially rainfall and factors controlling evapotranspiration.

Gaseous losses of nitrogen from cattle slurry and manure

Gaseous losses of N from cattle slurry and manure have been inferred from measurements of their N content following application [e.g., ref. 25]. Such indirect estimates suggest losses of 15 to 50% of the N applied and have usually been attributed to volatilisation of NH_3. However, denitrification is also expected to be important due to the association of organic substrates with the N applied and the effects of slurry on the physical properties of soil.

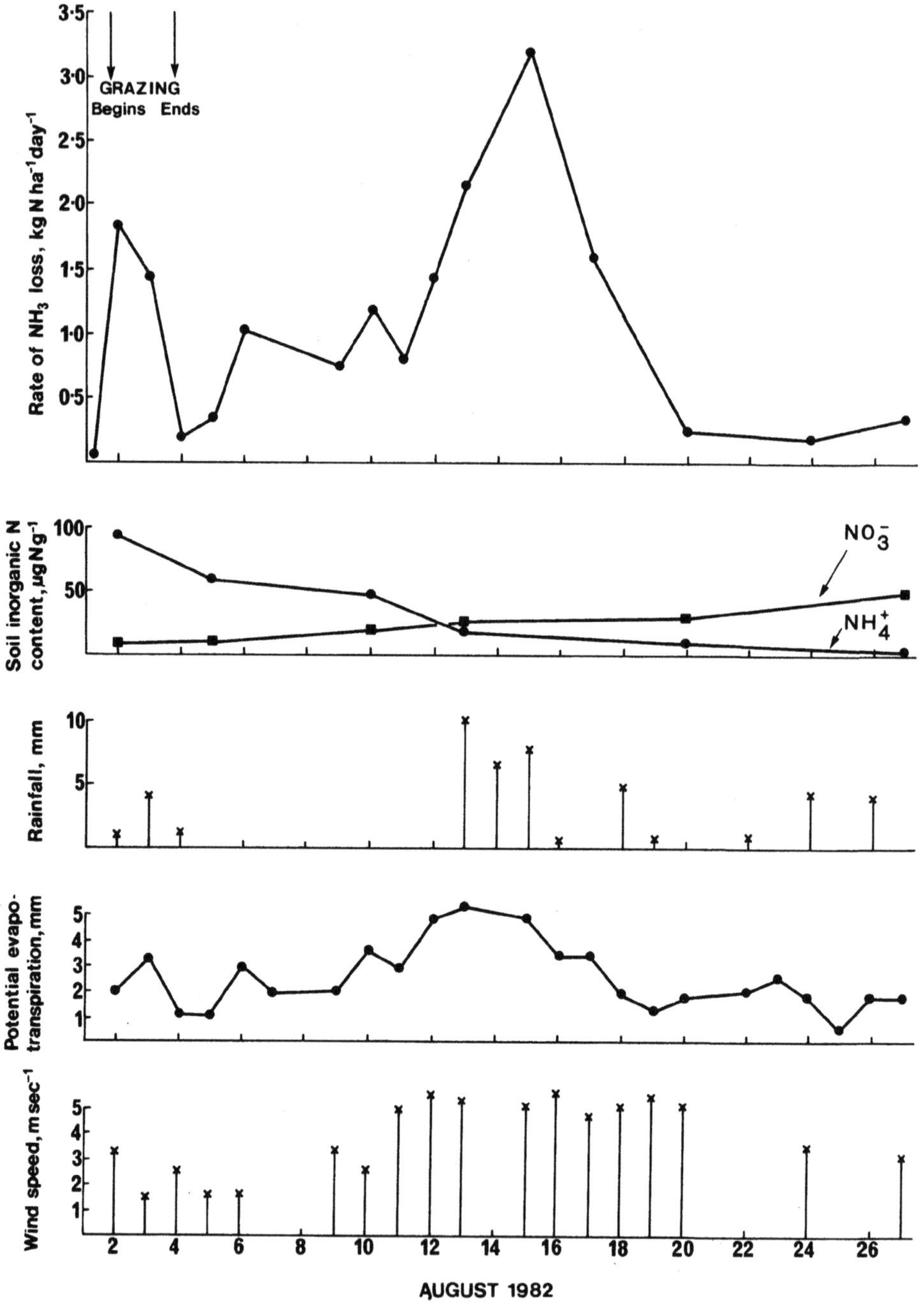

Figure 3. Rate of NH_3 loss from a perennial ryegrass sward on a freely drained loam during and after grazing by yearling steers in a 28-day rotation. The sward received 420 kg N ha^{-1} yr^{-1} as NH_4NO_3 in seven applications of 60 kg N ha^{-1}. Data for soil NH_4^+ and NO_3^- contents (0–5 cm) and meteorological conditions are provided for comparison. Wind speed was measured at a height of 3.5 m above the sward.

There are few data relating to denitrification loss of slurry-derived N. Burford *et al.* [7] reported that applications of cattle slurry (up to 550 t ha^{-1}) greatly extended the period during which N_2O was detected in the soil atmosphere. Oxygen concentrations in the soil atmosphere remained below 10% for at least four months. Rolston and Broadbent [30] reported that denitrification loss from NO_3^- applied to a fallow soil increased from 8 to 208 kg N ha^{-1} when manure was applied. The ratio of production of N_2:N_2O also increased from 2.7 to 20, reflecting a more intensely reduced environment in the manure-treated soil.

Similarly, Christensen [8] reported that rates of N_2O loss following application of cow slurry were as much as eight times greater than those following application of an equal amount of N as NH_4NO_3. In a recent study at Hurley (Ryden, unpublished), rates of denitrification in the range 0.01 to 0.05 kg N ha^{-1} day^{-1} were observed for a freely drained soil following irrigation of a sward treated with 100 kg slurry N ha^{-1} in late spring. Rates of loss were 5 to 10 times greater in the same soil into which slurry had been injected. It should be noted, however, that denitrification loss following application of slurry may not be large initially, as nitrification is pre-requisite to the formation of sufficient nitrate compatible with the enhanced potential for loss.

There are few direct measurements of NH_3 loss from slurry. Beauchamp *et al.* [4] have measured NH_3 losses equivalent to 20% of the N applied within 6 to 7 days of spreading dairy-shed slurry (325 kg N ha^{-1}) to a fallow soil in eastern Canada. The total loss of NH_3 was probably substantially higher as loss was continuing at rates between 0.3 and 0.7 kg N ha^{-1} hr^{-1} when measurements were terminated. Rates of loss as high as 2.9 kg N ha^{-1} hr^{-1} were observed during warm dry periods. More significant, however, was the fact that NH_3 continued to be lost at rates up to 0.3 kg N ha^{-1} hr^{-1} in cold wet weather, conditions that are usually assumed to minimize NH_3 loss. Nevertheless, the generally higher efficiencies of slurry N following applications in late winter and early spring [24] can probably be attributed to reduced NH_3 loss particularly when application occurs during periods of frequent rain.

In a recent study at Hurley (Ryden and Lockyer, unpublished), NH_3 loss was measured after surface application or injection of cattle slurry (40 t ha^{-1}) during warm dry weather in late spring. Following the surface application, NH_3 loss amounted to 30% of the total N applied (100 kg ha^{-1}) or approximately 80% of the content of inorganic N (largely NH_4^+) in the slurry. More than 95% of the total loss occurred within two days of application. A loss of $<2\%$ was observed for the injected treatment.

Prospects for reducing gaseous losses of nitrogen

From the limited amount of data currently available it is evident that substantial gaseous losses of N are associated with intensively managed grassland. Considerations of the flow of N in grassland [3] suggest that management to reduce losses will be most effectively applied 'down-stream' of the ruminant, i.e., through more effective re-cycling of the N returned in excreta. However, there is some possibility of decreasing gaseous losses that occur directly from fertilizer by adjustments to the time, amount and form of fertilizer applied. Most obvious are the attempts to reduce NH_3 volatilisation from urea by acidification, urease inhibitors, placement in the soil, or application prior to heavy rain [43]. With respect to denitrification, annual losses from NH_4NO_3 will be substantially decreased if that which occurs in the late winter/early spring can be avoided. At this time, a loss of 10 to 30% of the N applied is common and depends on soil conditions, particularly temperature in the surface soil in the first two weeks immediately following application [33]. This loss can be substantially reduced if a fertilizer containing or producing only NH_4^+-N is used [34].

Manipulation of the N flow downstream of the ruminant poses numerous problems. The most obvious point of intervention is through improved management of slurry and manure. Possible methods to reduce denitrification loss include the use of nitrification inhibitors, whilst addition of zeolites [34] prior or during application to land may decrease NH_3 loss. As discussed above, loss of NH_3 is greatly reduced when slurry is injected below rather than spread on the land surface [24]. Although this practice may also stimulate denitrification loss, increased residual value of the slurry N has been observed in some studies [45].

Prospects for reducing gaseous losses associated with grazing are more daunting. Swards could be irrigated following grazings in rotation to wash urine-N into the base of the sward thereby increasing contact with cation exchange sites in the soil surface. The addition of crushed natural zeolites to grazed swards may also achieve a reduction of NH_3 loss by increasing the cation exchange capacity at the base of the sward [34]. It is important to note, however, that any attempt to reduce the more immediate loss of N through volatilisation of NH_3 may only compound subsequent losses through denitrification and leaching of NO_3^- derived from the retained NH_4^+.

Conclusions and research needs

Recent developments in techniques to study gaseous losses of N, as they occur in the field, provide the opportunity to determine complete

Table 1. Inputs, outputs and storage of N in cut and grazed ryegrass swards on a freely drained loam. The swards were defoliated seven times each year at four-weekly intervals beginning in mid April. Fertilizer (60 kg NH_4NO_3-N/ha) was applied after each defoliation. The amounts of N moving into each sink or through each pathway of loss were determined independently

	Cut ryegrass	Grazed ryegrass
	kg N ha^{-1} yr^{-1}	
INPUTS		
Fertilizer	420	420
Atmosphere	15	15
	435	435
OUTPUTS		
Herbage or liveweight	300	29
Leaching	33	160
Denitrification	20	40
Ammonia volatilisation	0	120
STORAGE in soil organic matter	70	100
	423	449

balances for N in grazed grassland using a whole system approach. An illustration is given by the data in Table 1. In this study the flow of N into each sink or through each pathway of loss was determined independently. The balance between inputs and outputs of N was remarkably close. In the cut ryegrass sward, gaseous losses were small and much of the input of N was recovered in herbage or incorporated into soil organic matter. In contrast, in the grazed sward only 7% of the input of fertilizer N was recovered in liveweight of the yearling steers. Although some N was incorporated into the soil organic matter, about 38% was lost through a combination of NH_3 volatilisation and denitrification. A similar amount was lost through leaching.

Evaluation of the importance of and the factors affecting gaseous losses of N from grassland is at an early stage, reflecting the recent development of techniques suited to detailed field studies. It is essential that these techniques are applied more widely to study losses in different environments and to establish the major factors affecting the losses of N as they occur in the field. Only in this way will it be possible to identify and evaluate realistic strategies to minimize such losses of N which constitute a substantial amount of the unaccounted-for N in intensively managed temperate grassland [3].

References

1. Ball P R and Keeney D R (1983) Nitrogen losses from urine-affected areas of a New Zealand pasture, under contrasting seasonal conditions. In Smith J A and Hays V W, eds. Proceedings XIV International Grassland Congress (Lexington), pp 342–344. Boulder, Colorado: Westview

2. Ball P R, Keeney D R, Theobald P W and Nes P (1979) Nitrogen balance in urine-affected areas of a New Zealand pasture. Agronomy Journal 71, 309–314
3. Ball P R and Ryden J C (1984) Nitrogen relationships in intensively managed temperate grassland. Plant and Soil 76, 23–33
4. Beauchamp E G, Kidd G E and Thurtell G (1982) Ammonia volatilization from liquid dairy cattle manure in the field. Canadian Journal of Soil Science 62, 11–19
5. Blackmer A M, Bremner J M and Schmidt E L (1980) Production of nitrous oxide by ammonium-oxidising chemoautotrophic microorganisms in soil. Applied and Environmental Microbiology 40, 1060–1066
6. Bremner J M and Blackmer A M (1978) Nitrous oxide: emission from soils during nitrification of fertilizer nitrogen. Science 199, 295–297
7. Burford J R, Greenland D J and Pain B F (1976) Effects of heavy dressings of slurry and inorganic fertilizers applied to grassland on the composition of drainage waters and the soil atmosphere. In Agriculture and Water Quality, Technical Bulletin No. 32, pp 432–443, Ministry of Agriculture, Fisheries and Food. London: H.M.S.O.
8. Christensen S (1983) Nitrous oxide emission from a soil under permanent grass: seasonal and diurnal fluctuations as influenced by manuring and fertilization. Soil Biology and Biochemistry 15, 531–536
9. Conrad R and Seiler W (1980) Field measurements of the loss of fertilizer nitrogen into the atmosphere as nitrous oxide. Atmospheric Environment 14, 555–558
10. Denmead O T (1979) Chamber systems for measuring nitrous oxide emission from soils in the field. Soil Science Society of America Journal 43, 89–95
11. Denmead O T (1983) Micrometeorological methods for measuring gaseous losses of nitrogen in the field. In Freney J R and Simpson J R, eds. Gaseous loss of nitrogen from plant-soil systems, pp 133–157. The Hague: Martinus Nijhoff/Dr Junk
12. Denmead O T, Freney J R and Simpson J R (1976) A closed ammonia cycle within a plant canopy. Soil Biology and Biochemistry 8, 161–164
13. Denmead O T, Simpson J R and Freney J R (1974) Ammonia flux into the atmosphere from a grazed pasture. Science 185, 609–610
14. Denmead O T, Simpson J R and Freney J R (1977) Direct field measurement of ammonia emission after injection of anhydrous ammonia. Soil Science Society of America Journal 41, 1001–1004
15. Feagley S E and Hossner L R (1978) Ammonia volatilization reaction mechanisms between ammonium sulphate and carbonate systems. Soil Science Society of America Journal 42, 364–367
16. Galbally I E and Roy C R (1978) Loss of fixed nitrogen from soils by nitric oxide exhalation. Nature (London) 275, 734–735
17. Garwood E A and Ryden J C (1986) Nitrate loss through leaching and surface runoff from grassland: effects of water supply, soil type and management (this volume)
18. Gasser J K R (1964) Some factors affecting losses of ammonia from urea and ammonium sulphate. Journal of Soil Science 15, 258–271
19. Greenwood D J (1963) Nitrogen transformations and the distribution of oxygen in soil. Chemistry and Industry No 24, 799–803
20. Hargrove W L and Kissel D E (1979) Ammonia volatilization from surface applications of urea in the field and laboratory. Soil Science Society of America Journal 43, 359–363
21. Harper L A, Catchpoole V R and Vallis I (1983) Ammonia loss from fertilizer applied to tropical pastures. In Freney J R and Simpson J R, eds. Gaseous loss of nitrogen from plant-soil systems, pp 195–214. The Hague: Martinus Nijhoff/Dr Junk
22. Hutchinson G L and Mosier A R (1981) Improved soil cover method for field measurement of nitrous oxide fluxes. Soil Science Society of America Journal 45, 311–316
23. Hutchinson G L, Mosier A R and Andre C E (1982) Ammonia and amine emissions from a large cattle feedlot. Journal of Environmental Quality 11, 288–293
24. Kolenbrander G J (1981) Effect of injection of animal waste on ammonia losses by volatilization on arable land and grassland. In Brogan J C, ed. Nitrogen losses and surface runoff from landspreading of manures, pp 425–430. The Hague: Martinus Nijhoff/Dr Junk

25. Lauer D A, Bouldin D R and Klausner S D (1976) Ammonia volatilization from dairy manure spread on the soil surface. Journal of Environmental Quality 5, 134–141
26. Lemon E (1978) Nitrous oxide exchange at the land surface. In Nielsen D R and MacDonald J G, eds. Nitrogen in the environment, Vol. 1, pp 493–521. New York: Academic Press
27. Lockyer D R (1984) A system for the measurement of field losses of ammonia through volatilisation. Journal of the Science of Food and Agriculture 35, 837–848
28. MacDiarmid B N and Watkin B R (1972) The cattle dung patch. II. Effect of a dung patch on the chemical status of the soil and ammonia nitrogen losses from the patch. Journal of the British Grassland Society 27, 43–48
29. Rolston D E (1978) Application of gaseous-diffusion theory to measurement of denitrification. In Nielsen D R and MacDonald J G, eds. Nitrogen in the environment, Vol. 1, pp 309–335. New York: Academic Press
30. Rolston D E and Broadbent F E (1978) Field measurement of denitrification. EPA Technology Series EPA-600/2-77-233. Ada, Oklahoma: U.S. Environmental Protection Agency
31. Rolston D E, Fried M and Goldhamer D A (1976) Denitrification measured directly from nitrogen and nitrous oxide gas fluxes. Soil Science Society of America Journal 40, 259–266
32. Ryden J C (1981) N_2O exchange between a grassland soil and the atmosphere. Nature (London) 292, 235–237
33. Ryden J C (1983) Denitrification loss from a grassland soil in the field receiving different rates of nitrogen as ammonium nitrate. Journal of Soil Science 34, 355–366
34. Ryden J C (1984) Fertilizers for grassland. Chemistry and Industry No 18, 652–657
35. Ryden J C (1984) The flow of nitrogen in grassland. Proceedings of the Fertiliser Society (London) No 229, 44 pp
36. Ryden J C and Dawson K P (1982) Evaluation of the acetylene-inhibition technique for the measurement of denitrification in grassland soils. Journal of the Science of Food and Agriculture 33, 1197–1206
37. Ryden J C and Lockyer D R (1985) Evaluation of a system of wind tunnels for field studies of ammonia loss from grassland through volatilisation. Journal of the Science of Food and Agriculture 36, 781–788
38. Ryden J C, Lund L J and Focht D D (1978) Direct in-field measurement of nitrous oxide flux from soils. Soil Science Society of America Journal 42, 731–738
39. Ryden J C, Lund L J, Letey J and Focht D D (1979) Direct measurement of denitrification loss from soils: II. Development and application of field methods. Soil Science Society of America Journal 43, 110–118
40. Ryden J C and McNeill J E (1985) Application of the micrometeorological mass balance method to the determination of ammonia loss from a grazed sward. Journal of the Science of Food and Agriculture 36, 1297–1310
41. Ryden J C and Rolston D R (1983) The measurement of denitrification. In Freney J R and Simpson J R, eds. Gaseous loss of nitrogen from plant-soil systems, pp 91–132. The Hague: Martinus Nijhoff/Dr Junk
42. Siegel R S, Hauck R D and Kurtz L T (1981) Determination of $^{30}N_2$ and application to measurement of N_2 evolution during denitrification. Soil Science Society of America Journal 45, 68–74
43. Terman G L (1979) Volatilization of nitrogen as ammonia from surface-applied fertilizers, organic amendments and crop residues. Advances in Agronomy 31, 189–223
44. Tiedje J M, Sextone A J, Parkin T B, Revsbech N P and Shelton D R (1984) Anaerobic processes in soil. Plant and Soil 76, 197–212
45. Tunney H and Molloy S (1980) Research Report (Soils), pp 29–30. Dublin: An Foras Taluntais
46. Vallis I, Harper L A, Catchpoole V R and Weier K L (1982) Volatilization of ammonia from urine patches in a sub-tropical pasture. Australian Journal of Agricultural Research 33, 97–107
47. Webster C P and Dowdell R J (1982) Nitrous oxide emission from permanent grass swards. Journal of the Science of Food and Agriculture 33, 227–230

Relationships between soil mineral nitrogen content and denitrification, following application of slurry and calcium nitrate to grassland

GWEN M EGGINTON and KA SMITH
The Edinburgh School of Agriculture, West Mains Road, Edinburgh EH9 3JG, Scotland

Key words: denitrification, ammonium in soil, nitrate in soil, nitrogen in plants, nitrous oxide, gaseous diffusion

Abstract. Nitrogen (N) uptake, concentrations of mineral N remaining in the soil profile, and losses of N by denitrification were compared in a grassland soil fertilized with calcium nitrate or cattle slurry over a 2-year period. A far higher proportion of the mineral N in the soil in autumn was unaccounted for by plant uptake from the nitrate-treated than from the slurry-treated plot. Approximate estimates of the losses by denitrification indicated that these constituted a very significant fraction of the mineral N unaccounted for in the nitrate treatment. Spatial variability of soil nitrate concentrations could account for much of the observed variability in rates of denitrification.

It is widely accepted that the losses of nitrogen (N) from agricultural land by denitrification are significant, but available data are inadequate to make a quantitative estimate of these losses [9]. For N to be lost to the atmosphere by denitrification, it must first be present in the soil in the nitrate form. The N in manures is present largely as NH_4^+- and organic N and is thus protected from loss by denitrification until it has undergone mineralisation and/or nitrification. This paper reports the results of a comparison between the effect of one organic manure (cattle slurry) and that of inorganic N fertilizer, on the mineral N in the soil profile under a grass sward, and on the losses of N from the system. It also examines the relationship between the quantity of mineral N and the estimated extent of loss by denitrification.

Experimental methods

The field site was on a soil of the Macmerry series, classified as a stagnogleyic brown earth [1], and sown with perennial ryegrass (*Lolium perenne* L.), on the Edinburgh School of Agriculture's Langhill Farm, near Penicuik, Lothian. On 4 September, 1978, a 10 m x 8 m plot was treated with 100 kg N ha^{-1} as $Ca(NO_3)_2$; a similar plot was treated with 270 t ha^{-1} of cattle slurry containing 1230 kg N. A third plot was retained as an unfertilized control. The grass on the plots was cut and removed on three occasions over the next nine months, the dry matter (DM) yields were measured and the N content determined by Kjeldahl analysis [2]. Soil cores (75 mm diameter, eight from each plot)

H G van der Meer, J C Ryden and G C Ennik (eds) Nitrogen Fluxes in Intensive Grassland systems.
 ISBN 978-94-010-8454-3

Table 1. Dates of application and quantities of N applied as slurry and $Ca(NO_3)_2$ in the second experimental year

Date of application	Wt of slurry applied ($t\ ha^{-1}$)	N applied ($kg\ ha^{-1}$)	
		Slurry	$Ca(NO_3)_2$
16.7.79	54	59	100
20.8.79	80	74	100
29.10.79	95	71	100
15.4.80	142	94	100

were taken to 50 cm depth, and divided into 10 cm segments. Exchangeable ammonium and nitrate were extracted with 2*M* KCl, the NH_4^+ was determined by a phenol-nitroprusside colorimetric method [8], and the nitrate was determined by continuous-flow analysis [7].

In July 1979 the three plots were each split into three 8 m x 3 m subplots to which treatments (slurry, fertilizer nitrate and unfertilized control) were assigned at random. The dates of application and quantities applied are given in Table 1. The concentration of N in the slurry varied between applications, but was always much lower than in the single application the previous year. Again, herbage was harvested at intervals, and soil cores (two per plot) were taken. Plant DM and N content, and soil exchangeable NH_4^+ and NO_3^-, were determined as before.

In both years, the concentration of nitrous oxide (N_2O) was determined in samples of the soil atmosphere taken from 15, 30 and 45 cm depths, using the gas chromatographic techniques described by Dowdell *et al.* [4] and Smith [10]. The N_2O flux out of the soil was estimated from these concentrations and measured gaseous diffusion coefficients, using Fick's Law. A more detailed description of the experimental procedures is given in Egginton and Smith [5, 6].

Results and Discussion

The DM yields, N contents and quantities of N removed in herbage in year 1 are shown in Table 2. The uptake of N from the slurry-treated plot between 4.9.78 and 2.7.79, above that from the unfertilized control plot, was equivalent to about 3% of the quantity added. Less uptake was observed from the nitrate – treated plot than from the control, thus the recovery of nitrate was effectively zero. These results reflect short-term loss of N through NH_3 volatilisation from slurry, and subsequent winter losses through denitrification and leaching after relatively little growth and uptake following autumn application. The low recovery of slurry N also suggests little mineralisation of its content of organic N.

Table 2. Nitrogen removed in herbage from plots receiving slurry and $Ca(NO_3)_2$ in the first experimental year

Sampling date	Treatment	DM yield (kg ha^{-1})	N content (%)	N removed (kg ha^{-1})
6.11.78	S	500	3.47	17
	N	563	3.92	22
	C	188	3.17	6
11.6.79	S	1950	2.98	58
	N	1262	1.75	22
	C	1637	2.44	40
2.7.79	S	606	3.38	21
	N	525	1.70	9
	C	575	2.16	12
Total:	S	3056		96
	N	2350		53
	C	2400		58

S: slurry; N: nitrate; C: unfertilized control

Table 3. Exchangeable NH_4^+-N and NO_3^--N in soil profiles beneath the slurry – and nitrate-treated plots in the first experimental year

Sampling date	Depth (cm)	NH_4^+-N (mg kg^{-1})		NO_3^--N (mg kg^{-1})		
		S	C	S	N	C
25.9.78	0–10	2.2	3.7	1.4	93.5	48.0
	10–20	5.2	1.8	45.3	34.1	16.7
	20–30	7.2	0.8	31.8	21.7	3.4
	30–40	7.1	2.3	16.3	2.1	2.3
	40–50	28.5	1.9	7.3	0.6	2.1
12.6.79	0–10	1.9	2.2	12.4	9.5	7.1
	10–20	5.2	1.3	12.3	7.7	7.0
	20–30	0.8	1.8	6.8	3.0	3.9
	30–40	1.8	2.4	3.7	1.4	2.0
	40–50	0.8	0.4	2.3	1.0	1.2

S: slurry; N: nitrate; C: unfertilized control

The distribution of mineral N in the soil profile 3 weeks after treatments were applied in September 1978 showed several significant features (Table 3). The ammonium and nitrate concentrations were very low in the uppermost 10 cm of the slurry-treated plot, compared with the nitrate-fertilized and control plots, but nitrate was higher in the 10–50 cm zone and ammonium increased with depth, the highest value occurring at 40–50 cm. This suggests that the slurry (containing water equivalent to 27 mm of rainfall) moved rapidly down the profile, leaching the nitrate initially present from the uppermost 10 cm. Most of the slurry N (NH_4^+ initially present or mineralised from organic constituents) had been nitrified at depths shallower than 40 cm. In

Table 4. Total quantities of NH_4^+-N and NO_3^--N in uppermost 50 cm of soil profile beneath the slurry- and nitrate-treated plots in the first experimental year

Sampling date	NH_4^+-N (kg ha^{-1})		NO_3^--N (kg ha^{-1})		
	S	C	S	N	C
25.9.78	65	14	133	198	94
12.6.79	14	11	49	29	28

S: slurry; N: nitrate; C: unfertilized control

Table 5. Reductions in mineral N content of the soil profile N, removed in harvested herbage, and the quantity of N unaccounted for by plant uptake, between 25 September 1978 and 12 June 1979

Treatment	Nitrogen (kg ha^{-1})		
	Reduction in mineral N	Removed in harvested herbage	Unaccounted for
Nitrate	169	44	125
Slurry	135	76	59
Control	69	46	23

contrast, slurry N present at 40–50 cm, which presumably reached this depth by percolation through the macropores, was still largely in the ammonium form. In the $Ca(NO_3)_2$ plot, nitrate had moved to a depth of 30 cm (Table 3).

On the first sampling date only 90 kg inorganic N per ha were recovered, relative to the control, in the slurry-treated plot, i.e. 7% of the total N applied (Table 4). For the nitrate-fertilized plot, even though 16 kg more N were taken up than from the control, the increase in inorganic N in the top 50 cm was 104 kg, compared with the 100 kg added. Thus there was some plot-to-plot variability in the quantity of inorganic N released from non-fertilizer sources, but clearly no obvious loss of applied nitrate by leaching or denitrification had occurred by this first sampling date. The following summer, there was little evidence of increased ammonium in the slurry-treated profile, but nitrate concentrations were higher than in the other treatments at all depths, indicating a prolonged release of N from the added organic matter (Tables 3 and 4). Although the nitrate-treated plot had by far the greatest total nitrate content in the profile on 25 September, this had dropped to the level of the control by the following summer.

There was no obvious physical damage to the sward following the application of slurry, and the quantities of dry matter and N removed as herbage were greatest from the slurry-treated plot. However, the reduction in mineral N between 25 September 1978 and 12 June 1979, unaccounted for by plant uptake, was by far the greatest where nitrate

Table 6. Nitrogen removed in herbage from plots receiving N as slurry and calcium nitrate, in the second experimental year (total of 6 cuts between 9 August 1979 and 19 June 1980)

Treatment	DM yield (kg ha^{-1})	N removed (kg ha^{-1})
Slurry	5560	143
Nitrate	7140	226
Control	3090	85

Table 7. Total quantities of NH_4^+-N and NO_3^--N in the uppermost 40 cm of soil profiles beneath the slurry- and nitrate-treated plots in the second experimental year

Sampling date	NH_4^+-N (kg ha^{-1})		NO_3^--N (kg ha^{-1})		
	S	C	S	N	C
9.8.79	2	2	4	12	10
1.10.79	2	2	12	18	15
10.1.80	6	5	39	34	30
15.4.80	4	4	9	13	8
22.7.80	ND	ND	14	12	10

S: slurry; N: nitrate; C: unfertilized control; ND: not determined

had been applied (Table 5). This conclusion can be drawn even though NH_4^+-N was not measured in the nitrate plot, because it is assumed that it was of the same order as that found in the control plot prior to fertilizer application.

The quantity of N lost over the first season as N_2O was estimated to be in the range 1.2–5.3 and 2.6–11.2 kg ha^{-1} in the slurry- and nitrate-treated plots, respectively [5]. When the acetylene-inhibition method was used in a later experiment on the same site, the ratio of total loss of N as (N_2 + N_2O) to that as N_2O alone in the nitrate-treated plot ranged from 1.9 in the autumn to 8.3 in the spring [6]. If these factors are applied to the N_2O losses observed in 1978–79, then the total loss by denitrification in the nitrate plot was in the range 6–91 kg N ha^{-1}, with a mean of 36 kg N ha^{-1}.

It is recognised that this procedure can only give a qualitative guide to the likely losses by denitrification, rather than a quantitative measure. Nevertheless, the data clearly suggest that a significant fraction of the nitrate-N unaccounted for in the nitrate-treated plot was lost by denitrification. The nitrate-N unaccounted for in the slurry-treated plot was intermediate between that in the nitrate-treated plot and the control. A significant part of this was also likely to have been lost by denitrification; however, the unknown effects of the increased O_2 demand on the extent of anaerobic conditions in the slurry-treated soil prevent a similar calculation to that made for the nitrate-treated plot.

In the second year, there was greater growth of herbage and consequently more N was removed in each harvest (Table 6). The mineral N concentrations remaining in the soil profile were lower than in the

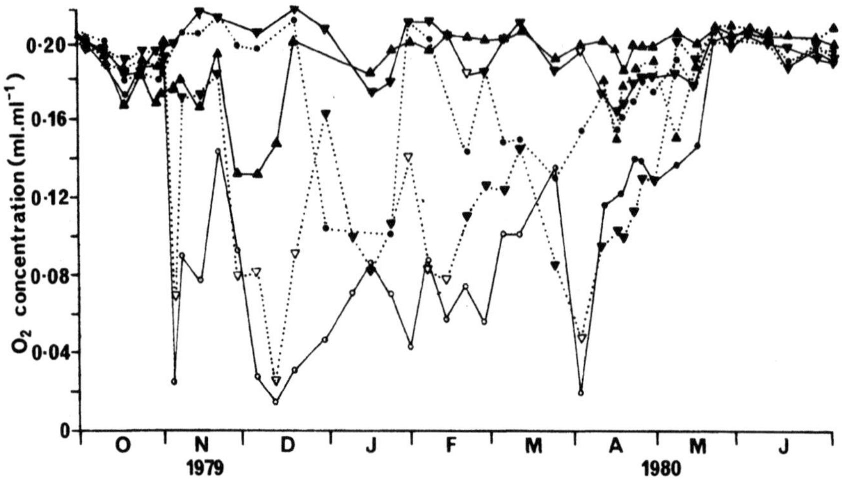

Figure 1. O_2 concentrations in gas or water samples from replicate probes in nitrate-fertilized plots (45 cm depth). O_2 concentrations in solution converted to equivalent gas phase concentrations with which they would have been in equilibrium [11]. ●, ▲, ▼ : gas samples; ○, ▽ : water samples.

previous year (Table 7). In this second season, denitrification was also lower; the calculated losses of N as N_2O were 0.4–1.8, 0.5–2, and 0.2–0.7 kg ha^{-1} for the slurry, nitrate and control plots, respectively. It cannot be said with certainty that the lower denitrification was due to the lower mineral N concentrations and not to the generally drier conditions which prevailed in the second year. However, the period of most rapid denitrification occurred when the concentrations of nitrate-N in the soil profile were highest: from November to January. Further work is required to determine the seasonal variation in the relationship between soil mineral N content and losses by denitrification.

In both years, the calculated losses of N by denitrification covered a wide range of values for each treatment. It is unlikely that this situation can be avoided, given the inherently variable nature of the denitrification process. This is well illustrated in Figs. 1 and 2, which show the variability between replicate soil atmosphere sampling probes in O_2 and N_2O concentrations. The spatial variability of O_2 was greatly influenced by the random occurrence of waterlogging. Those sampling probes which consistently exhibited the lowest O_2 concentrations were those which were intermittently or continuously waterlogged during the winter months (Fig. 1). Concentrations of N_2O were also consistently higher at some sampling locations, with a 5–10 fold difference between highest and lowest (Fig. 2). These results are in keeping with the concept that denitrification will only occur at those points in the

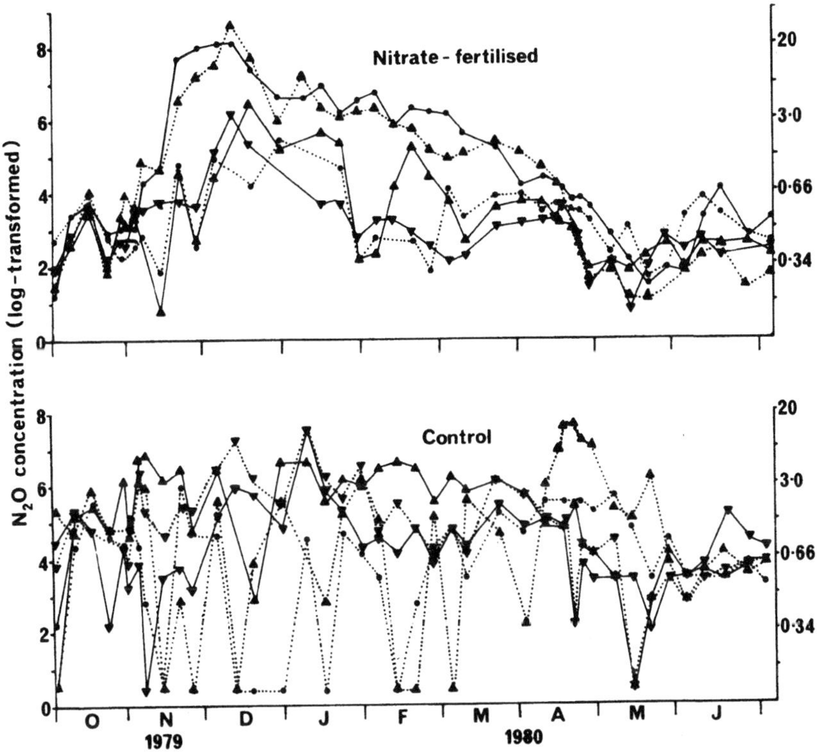

Figure 2. N_2O concentrations in samples from replicate probes in nitrate-fertilized and control plots (45 cm depth). Transformed to log-values by the relationship $X' = 5 + \ln(X - 0.29)$, where X = measured N_2O concentration (ml ml^{-1} $\times 10^6$), X' = transformed value, 0.29 = ambient N_2O concentration (ml ml^{-1} $\times 10^6$). The constant, 5, is included to avoid negative values.

soil where O_2 has virtually disappeared. Slight differences in O_2 diffusion rates can result in some locations being just anaerobic and therefore allowing denitrification to proceed, while others remain just aerobic and do not denitrify at all.

Figure 3 shows that when all measurements are pooled, there is a clear inverse relationship between N_2O and O_2 concentrations in the soil atmosphere, but the figure also confirms the variability of the relationship. At each O_2 concentration the range of N_2O values generally covered two orders of magnitude. The trend shown in Fig. 3 is consistent with that reported by Dowdell and Smith [3] for an arable clay soil in the English Midlands. Since nitrate concentrations varied with depth and sampling occasion by up to two orders of magnitude (Table 3), and also showed wide variations between replicate cores, it is to be expected that comparable variations in the rate of evolution of N_2O would occur

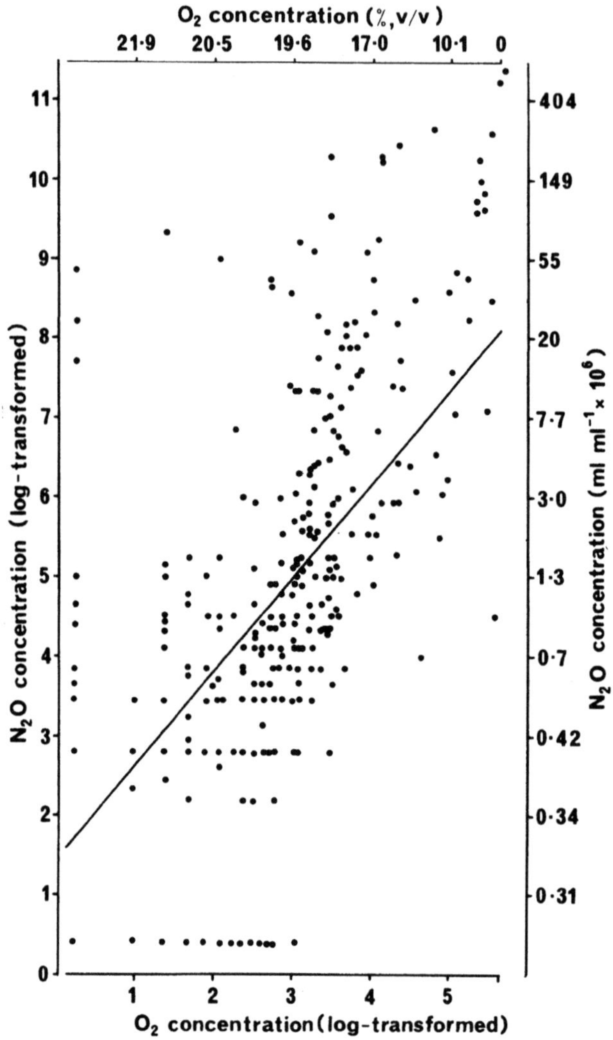

Figure 3. Scatter diagram of N_2O concentrations against corresponding ones of O_2 (nitrate-fertilized plots, 30 cm depth). N_2O concentrations log-transformed as for Fig. 2. O_2 concentrations transformed by the relationship $Y' = 2.6 + \ln(21.09 - Y)$, where Y = measured O_2 concentration (ml ml^{-1} $\times 10^2$).

(provided that other factors, e.g. organic substrates, did not become limiting). This could account for much of the variation shown in Fig. 3.

Acknowledgement The authors thank W. MacDougall, R.S. Howard and I.S. Crichton for technical assistance.

References

1. Avery B W (1980) Soil classification for England and Wales (higher categories). Monograph 14. Harpenden, UK: Soil Survey
2. Chapman H D and Pratt P F (1961) Methods of analysis for soils, plants and water. University of California
3. Dowdell R J and Smith K A (1974) Field studies of the soil atmosphere II. Occurrence of nitrous oxide. Journal of Soil Science 25, 231–238
4. Dowdell R J, Smith K A, Crees R and Restall S W F (1972) Field studies of ethylene in the soil atmosphere – equipment and preliminary results. Soil Biology and Biochemistry 4, 325–331
5. Egginton G M and Smith K A (1986) Nitrous oxide emission from a grassland soil fertilised with slurry and calcium nitrate. Journal of Soil Science 37, 59–67
6. Egginton G M and Smith K A (1986) Losses of nitrogen by denitrification from a grassland soil fertilised with cattle slurry and calcium nitrate. Journal of Soil Science 37, 69–80
7. Henriksen A and Selmer-Olsen A R (1970) Automatic methods for determining nitrate and nitrite in water and soil extracts. Analyst 95, 514–518
8. Kempers A J (1974) Determination of sub-micro quantities of ammonium and nitrates in soils with phenol, sodium nitroprusside and hypochlorite. Geoderma 12, 201–206
9. Royal Society (1983) The nitrogen cycle of the United Kingdom, a Study Group Report. London, UK: The Royal Society
10. Smith K A (1983) Gas chromatographic analysis of the soil atmosphere. In Smith K A, ed. Soil analysis, pp 407–454. New York, USA: Marcel Dekker
11. Smith K A and Russell R S (1969) Occurrence of ethylene, and its significance in anaerobic soil. Nature 222, 769–771

Losses of nitrogen from intensive grassland systems by leaching and surface runoff

JHAM STEENVOORDEN, H FONCK and HP OOSTEROM
Institute for Land and Water Management Research (ICW), P.O. Box 35, 6700 AA Wageningen, The Netherlands

Key words: nitrate leaching, nitrogen in surface runoff, methods to assess losses, grassland fertilization, reduction of nitrogen losses

Abstract. Evidence is growing that grassland intensification increases the risk of nitrogen (N) pollution of groundwater aquifers and surface water. For water quality management purposes, it is necessary to quantify the relationship between rates and times of fertilizer applications and resulting N loading of waters. Different techniques for quantification are available and the merits and drawbacks of some of these techniques are discussed. A strong correlation exists between nitrate leaching losses to groundwater and the level of N fertilization. The most important factors which influence nitrate leaching at a particular level of fertilization are soil type, soil use and hydrological situation. The highest nitrate concentrations are usually observed below grazed grassland. Surface runoff waters from fertilized grassland soils can contain high concentrations of ammonium and organic N, not only in hilly areas but also on land with gentle slopes. The amount of animal manure and the time interval between spreading and a surface runoff event are principle factors influencing concentrations of N in surface runoff. Reduction of N losses via leaching and surface runoff can be achieved by measures which influence the amount of mineral N in and on the soil and the hydrological situation.

On intensively managed grassland farms, large amounts of nitrogen (N) are necessary to maintain a high level of grass production and to produce grass of good quality. Evidence is growing that grassland intensification increases the risk of nitrate pollution of groundwater aquifers and surface waters [7, 9, 15].

Nitrate pollution of groundwater resources is of increasing concern with respect to both the supply of drinking water and the conservation of the natural environment. In the Netherlands increased nitrate concentrations have been measured in groundwater sampled at depths up to seventy metres [1]. It is not surprising therefore that in some regions nitrate concentrations in raw drinking water pumped from groundwater aquifers have increased rapidly in recent years (Fig. 1).

In order to protect the future supply of drinking water and natural ecosystems, it is necessary to quantify the contribution of all possible sources to the nitrate load of ground and surface waters. Some methods for assessing N losses via leaching and surface runoff will be discussed. The consequences of agricultural intensification especially on grassland farms for N losses to ground and surface waters will be assessed and possible measures to reduce such losses will be discussed.

H G van der Meer, J C Ryden and G C Ennik (eds) Nitrogen Fluxes in Intensive Grassland systems.
 ISBN 978-94-010-8454-3

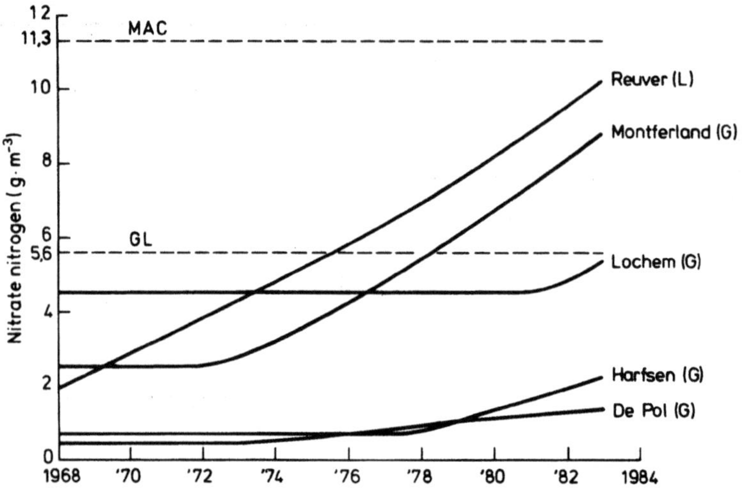

Figure 1. Nitrate concentration in raw drinking water pumped from groundwater aquifers in the provinces of Gelderland (G) and Limburg (L). The maximum admissible concentration (MAC) and the guide level (GL) from the EC directive for nitrate in drinking water have been indicated.

Methods to assess nitrogen losses to ground and surface waters

Quantification of N losses to ground and surface waters requires data about the amount of drainage or runoff water and its chemical composition. Different techniques are available to quantify water movement and accompanying N loss. Each technique has its own merits depending on the problem to be assessed and technical constraints imposed by the field site in question. The range of techniques available is summarised in Table 1. The principle difference between techniques relates to the depth of sampling involved.

Analysis of soil mineral N (technique 1) in the upper metre in autumn gives only an estimate of the potential loss of nitrate through leaching. The actual amount of N leached also depends on the amount of excess precipitation and changes in the mineral N content of soil brought about by microbiological processes which occur between the time of soil sampling and the end of the leaching season.

Soil water samples can be collected from below the root zone by using ceramic cups (technique 2). The leachate is sampled by creating a vacuum in the cup which exceeds soil water tension in the surrounding soil. The ceramic cups are connected to a vessel on the soil surface, under vacuum during sampling, in which the leachate is collected. For experiments on cut grassland with an even distribution of fertilizers and animal manures, four cups per field are sufficient to produce reliable data (Table 2).

Table 1. Techniques for quantification of N losses from soils to ground and surface waters[a]

Technique	Water quantity	N concentration	Total load
Leaching			
1. Soil mineral N	–	–	±
2. Soil moisture analysis at roughly 1 meter below soil surface	–	+	–
3. Borehole water from shallow groundwater	–	+	–
4. Lysimeter effluent	+	+	+
5. Drainage effluent	+	+	+
Surface runoff			
6. Gutter	+	+	+

[a] +, direct measurement possible
–, direct measurement not possible
±, provides only an estimate

Table 2. Comparison of nitrate N concentrations in leachate ($g\,m^{-3}$) from two sampling sets in one experimental field, each set consisting of four ceramic cups at one metre below soil surface. The experiment is on cut grassland; cattle slurry was injected in early spring. The water samples were taken in the drainage period after the growing season in 1984

N application ($kg\,ha^{-1}\,yr^{-1}$)		Nitrate N concentration in leachate ($g\,m^{-3}$)	
slurry	mineral fertilizer	set 1	set 2
143	600	36	35
143	400	13	9
143	0	2.5	2.3
283	600	83	87
283	400	25	25
283	0	3.2	5.5

At sites with a shallow groundwater table, leachate samples can also be collected from borehole water (technique 3). When information is required on N leaching from the preceding growing season, the borehole water should be sampled when water movement through the profile ceases. This is normally the beginning of April in the Netherlands. The depth of the borehole should approximate the percolation depth of excess precipitation during the wet period. Percolation depth can be estimated from soil water balance calculations and physical data for the soil. On grazed fields attention should be paid to the number of boreholes per field as a wide variation in concentration might occur (Table 3). In this case, more than 24 boreholes may be required to obtain reliable results. The high nitrate concentrations below these fields may partly be caused by the low grass production in the very dry summer of 1976.

Table 3. Variation of nitrate N concentrations in the upper metre of the groundwater below grazed fields. Results are based on separate analyses of groundwater from 15 or 24 boreholes per field early in the spring of 1977

$\bar{x}$ = mean nitrate N concentration

Sx = standard deviation of the 15 or 24 analyses

$S\bar{x}$ = standard deviation of the mean ($Sx/\sqrt{n}$)

		Nitrate N ($g\,m^{-3}$)			
Field	Number of boreholes	$\bar{x}$	Sx	$S\bar{x}$	$S\bar{x}$ (% of $\bar{x}$)
1	24	80	47	9.6	12.0
2	24	111	57	11.6	10.5
3	24	95	56	11.4	12.0
4	15	137	58	15.0	10.9
5	15	137	68	17.6	12.8
6	15	65	48	12.4	19.1
7	15	80	27	7.0	8.7

Results from lysimeter experiments (technique 4) will only be applicable to field situations if the soil moisture regime within the root zone is the same as that in the undisturbed soil in the field. Comparability is often poor due to the fixed groundwater level in lysimeters, physical disturbance during installation of lysimeters and absence of surface runoff in lysimeter experiments. It is impossible to assess the effects of grazing in lysimeter studies.

Samples collected from field drains (technique 5) can also reflect the influence of soil use. It should be noted, however, that field drain effluents are a mixture of water with different residence times in the subsoil. Precipitation on soil immediately above a drain may be discharged in the same year. In contrast, the residence time of water in subsoil between two drains may be several years depending on the hydrological situation [4]. Consequently, the concentration of nitrate in field drain effluents may reflect not only the influence of land use during the preceding growing season, but also that in previous years as well as the effects of microbiological processes occurring during transport through the subsoil.

The disadvantages of water quality measurements using lysimeters and drains are balanced to some extent by the ease of water quantity measurements. In experiments using soil moisture and borehole techniques, water fluxes can only be estimated from soil water balance calculations. The choice between the available techniques will be influenced not only by considerations of the reliability of estimates of water quality or quantity, but also by other factors. The borehole technique, which results in considerable structural damage to the soil, is not applicable to small experimental plots if the experiment is to be continued for some years. Soil water analysis (technique 2) requires

Table 4. Nitrate concentrations in the upper metre of groundwater below grazed grassland on a sandy soil with additional slurry dressings [11]. Sampling via borehole technique

Farm number	Mineral N fertilizer dose ($kg\ ha^{-1}\ yr^{-1}$)	Additional slurry ($m^3\ ha^{-1}\ yr^{-1}$)	Groundwater table in spring (m below soil surface)	Nitrate N concentration in groundwater ($g\ m^{-3}$)
1	250	45 m^3 cattle slurry	1.0	18
1	250		0.75	0.5
2	505	10 m^3 cattle + 18 m^3 pig slurry	1.0	90
2	505		0.5	0.4

frequent sampling during the drainage season and is more labour intensive than the borehole technique.

Surface runoff is the excess precipitation which is transported over the soil surface to surface waters. Measurement of the amount of surface runoff and reliable sampling of its nutrient content can only be achieved by installation of a gutter at the edge of a field (technique 6).

Nitrogen losses via leaching

Leaching takes place only when precipitation exceeds evapotranspiration. Normally the growing period is characterized by a shortage of precipitation so that there is hardly any leaching. Leaching of N can be expected therefore at the end of the growing season when mineral N is frequently present in the soil profile. For cut grassland on both sandy and clay soils the amount of mineral N in the soil profile at the end of the growing season increases when rates of fertilizer N application exceed 350 $kg\ ha^{-1}\ yr^{-1}$ [14]. Leaching of N takes place mainly in the form of nitrate as adsorption onto soil particles and nitrification normally prevent leaching of ammonium. However, application of large amounts of animal manure in autumn will increase the risk of ammonium leaching. For example, 300 t of pig slurry per ha applied in November to a light sandy soil with a groundwater table at one metre below soil surface resulted in a concentration of ammonium plus organic N of 6.5 $g\ m^{-3}$ in the upper groundwater. At an application rate of 60 $t\ ha^{-1}\ yr^{-1}$ a N concentration of only 1.5 $g\ m^{-3}$ was found [12]. The organic N will gradually be converted to ammonium during transport to greater depths.

Nitrate leaching is influenced by many factors including land use, amount of fertilizer N applied, the type of fertilizer, soil type and water regime. Nitrate concentration in the upper groundwater is also influenced by the groundwater table as illustrated by the data in Table 4. Big differences were observed in the concentration of nitrate in

Table 5. Additional leaching of nitrate N per increment of mineral fertilizer N applied to cut grassland on sandy soil with a groundwater depth of at least one metre. Data are expressed as a percentage of the extra applied N and for a drainage volume of approximately 300 mm yr^{-1}

Increment of fertilizer N ($kg\ ha^{-1}\ yr^{-1}$)	N leaching (% of extra N)
0–200	0–5
200–300	15
300–400	35
400–500	45
500–600	60

groundwater under different fields of the same farm even though fields were managed in the same way. It is likely that a higher denitrification rate is responsible for the lower nitrate concentrations below fields with a shallower groundwater table. The importance of groundwater table in nitrate leaching has been confirmed in an experiment in which three lysimeters were installed with a depth of 3.3 m and a maximum groundwater depth in springtime of 0.5, 1.0 and 1.5 m. When 300 t of pig slurry per ha per year (1690 kg N) were applied in autumn, nitrate leaching losses amounted 5, 13 and 32% of the total applied N for the 0.5, 1.0 and 1.5 m groundwater levels, respectively. One third of the difference is caused by a higher uptake of N by the sward. The remainder might be explained by the more favourable conditions for denitrification in the root zone at a higher groundwater table. The results indicate that the groundwater table plays an important role in reducing leaching losses when its level below soil surface is less than 1.5 m.

Sandy soils show higher nitrate losses than clay soils. An application of mineral fertilizer N to grassland of 380 kg $ha^{-1}\ yr^{-1}$ leads to a N loss of approximately 20 kg $ha^{-1}\ yr^{-1}$ for clay soils [3, 9] and of 70 kg $ha^{-1}\ yr^{-1}$ for sandy soils [10]. This difference in nitrate leaching is partly due to a higher crop uptake on clay soils. Two other factors also lead to lower leaching losses from clay soils. Their finer structure is more favourable for denitrification and clay soils have a higher soil water content at field capacity. The latter results in a lower amount of leachate at the same amount of excess precipitation [16].

Results from leaching experiments at different rates of N fertilization carried out on lysimeters and drained fields have been compiled by Kolenbrander [10]. In general, these experiments have been conducted at sites where the groundwater table is at a depth of at least 1 m. Nitrate leaching below cut grassland on sandy soils is minimal when the amount of N applied is less than 200 kg $ha^{-1}\ yr^{-1}$ and amounts to only 0–5% of the total input of N. At higher rates of application, leaching losses increase rapidly (Fig. 2). The proportion of N leached for each increment in fertilizer application, is given in Table 5. For N

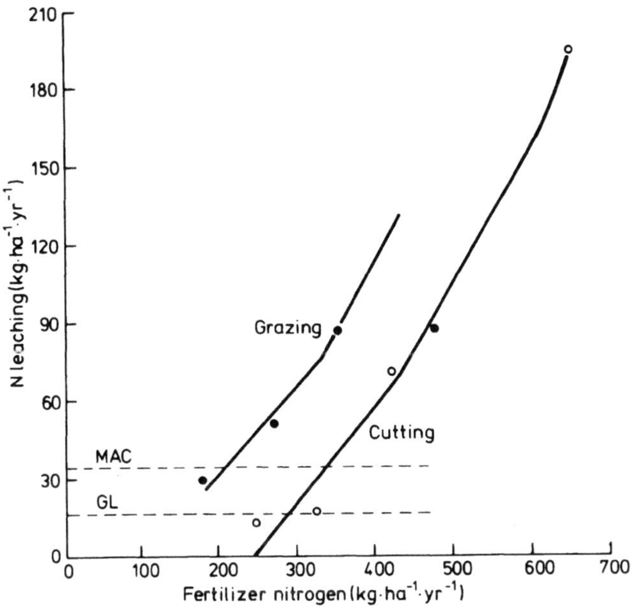

Figure 2. Relationship between applied artificial fertilizer N and nitrate in the leachate from cut (○) [10] and grazed (●) grassland [13] on sandy soils at a groundwater level of at least one metre below the soil surface. Leaching of N has been calculated for a drainage volume of 300 mm per year. The amounts of N which correspond to the maximum admissible concentration (MAC) and the guide level (GL) from the EC directive for nitrate in drinking water have been indicated.

applications greater than 400 kg ha^{-1} yr^{-1}, more than 50% of the additional fertilizer N will leach. The maximum admissible nitrate concentration for drinking water (11.3 g m^{-3} as N) is reached at a N fertilization level of 350 to 400 kg ha^{-1} yr^{-1} on cut grassland.

Studies on grazed grassland indicate that nitrate leaching is higher than that from cut grassland for the same level of application of fertilizer N (Fig. 2). For cattle farms on sandy soils receiving mineral N fertilizer at rates between 200 and 450 kg ha^{-1} yr^{-1}, nitrate N concentrations in leachate are approximately 10–20 g m^{-3} higher than for cut grassland [13]. The stocking density ranged from 2.5 to 3.5 cattle units per ha (1 unit = 1 milking cow). All slurry N produced by the stock was spread on the land. At a drainage water production of 300 mm yr^{-1} total N losses show an increase of 30–60 kg ha^{-1} yr^{-1}, compared to the cut swards, which roughly amounts to 10–20% of total N excreted by the animals. Separate analyses of water collected from boreholes on grazed grassland show a wide variation in the nitrate concentration (Table 3), probably reflecting the irregular distribution of faeces and urine during grazing.

Table 6. Input of N via mineral fertilizers and cattle slurry in spring and losses of nitrate N via leaching for grazed grassland on a sandy soil. The results for 1977 and 1978 are the averages for 9 and 8 fields, respectively [5, 6]

		1977	1978
Artificial fertilizer	(kg N ha^{-1})	460	400
Cattle slurry	(kg N ha^{-1})	40	130
Soil mineral N (0–100 cm depth)	(kg N ha^{-1})	–	280
N leaching	(kg N ha^{-1})	77	225
Amount of leachate	(mm)	214	375
N concentration	(g m^{-3})	36	60

Leaching losses from grazed grassland can vary substantially from year to year [5, 6]. In one study in the Netherlands, measurements were made of the inputs of N in mineral fertilizer and cattle slurry to each field on a farm on a sandy soil. The amount of N leached in the winter periods following the growing seasons in 1977 and 1978 differed by a factor 3 (Table 6). The amount of N leached in 1978 amounted 80% of the mineral N present in the soil profile before the drainage season commenced.

A recent development in grassland fertilization is the application of slurry via injection. Studies of the consequences of injection as opposed to surface spreading on a loamy sand were initiated in 1980. The amounts of nitrate leached after the growing seasons of 1982 and 1983 for fields which received different quantities of mineral fertilizer and slurry are given in Fig. 3. The highest rate of application of slurry was 40 t ha^{-1} yr^{-1} for surface application and 80 t ha^{-1} yr^{-1} for injection. Application of fertilizer N ranged from 0 up to 600 kg ha^{-1} yr^{-1} [19]. The relationship between available mineral N in the soil and amount of nitrate leached for injected slurry is essentially the same as that for surface applied slurry. The amount of available mineral N was calculated as the sum of mineral N given as fertilizer and that originating from the applied slurry. For slurry, only a part of the total N applied is assumed to contribute to the available mineral N in the soil. Based on grass production following surface application of slurry, only 25% of the total N applied was assumed to contribute to the mineral N in the soil. Following injection, 75% of the total N in slurry was assumed to contribute to the available mineral N in the soil. This is the mineral N content of the slurry at the time of injection (50% of total N) plus the easily mineralizable fraction (25%) [2].

It can be concluded that injection of slurry on grassland does not increase the risk of nitrate leaching when the rates of N application are adjusted to take account of the higher availability of mineral N for grass production. The somewhat lower N leaching in this experiment compared with Fig. 2 may be caused by the fine structure of the loamy sand which promotes denitrification.

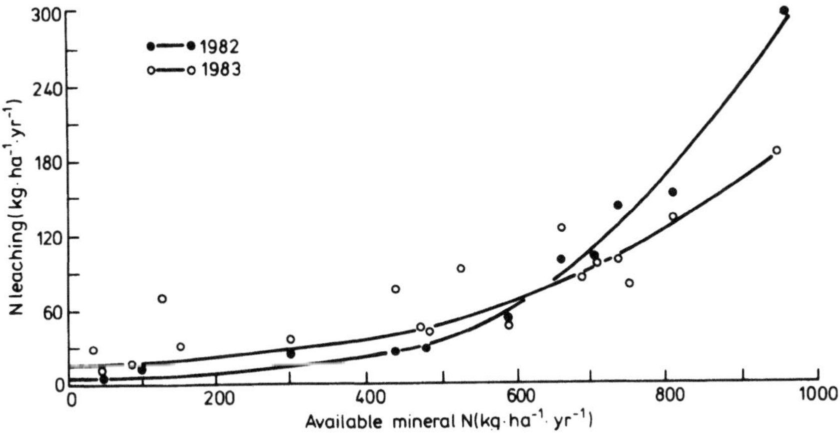

Figure 3. Relationship between nitrate leaching and total available mineral N in the soil from artificial fertilizer and slurry applications to cut grassland on a loamy sand. Of the total N in surface applications of slurry, 25% has been assumed to contribute to available mineral N in the soil. A value of 75% is assumed for injected slurry. Nitrate concentrations were determined in water extracted from a depth of one metre using four ceramic cups per field. Water samples have been collected in the drainage periods following the growing seasons of 1982 (•) and 1983 (○).

Surface runoff

Surface runoff or overland flow is a hydrological phenomenon which is influenced by the physical characteristics of the soil, ease of water drainage through the profile and precipitation intensity and its duration and distribution through the year. These interacting factors make surface runoff a complex process. Precipitation which cannot be absorbed by the soil is transported over the soil surface. In sloping areas this may cause soil erosion, but in the Netherlands soil erosion in general is of little importance. Although surface runoff occurs principally in a period (November to February) when no mineral fertilizer is applied, substantial amounts of animal slurries are spread at this time. The risk of surface runoff is especially high for soils with a high groundwater table which are badly drained. In north west Europe, these soils are usually under grass.

Concentrations of N in surface runoff water are influenced by amounts of slurry applied, time elapsed after slurry spreading before the next surface runoff event and amount of precipitation infiltrated in the period after slurry spreading. Field experiments in Ireland have illustrated that following an application of pig slurry equivalent to 3.7 t of dry matter per ha, N concentrations of dissolved ammonium can exceed 150 g m^{-3} when runoff occurs within two days of slurry application [17]. The longer the time interval between application and runoff,

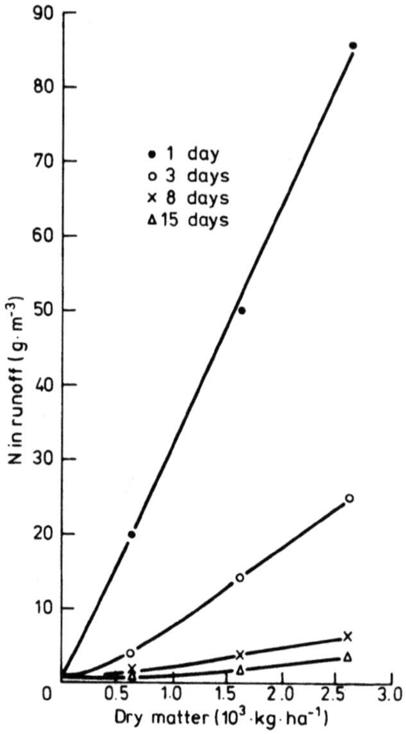

Figure 4. Concentration of dissolved Kjeldahl-N (ammonium plus organic N) in surface runoff waters from grassland fertilized with 0, 0.7, 1.6 and 2.6 t of dry matter per ha as cattle slurry. The N content of the slurry was 48 g per kg dry matter. Each field was irrigated four times (day 1, 3, 8 and 15 after application) to simulate 25 mm rainfall per irrigation event.

the lower the concentration of N in the runoff. After a five-day interval, the N concentration had fallen below 25 $g\,m^{-3}$. The effect of rate of slurry application on N concentrations in surface runoff has also been studied in a field experiment on grassland in the Netherlands [18]. Four surface runoff events have been created on each field by simulating 25 mm rainfall 1, 3, 8 and 15 days after spreading slurry at three different rates. The rate of slurry application had a substantial effect on dissolved N concentrations in surface runoff (Fig. 4). For every ton of dry matter applied per ha, the concentration of N in runoff increased by more than 30 $g\,m^{-3}$ when a runoff event was induced 1 day after slurry spreading. Three days after slurry spreading and following a cumulative artificial rainfall of 50 mm, the increase in N concentration per ton of dry matter decreased to 10 $g\,m^{-3}$. The lower concentrations of N observed for the longer time intervals after spreading can be explained by the lower amounts of slurry N on the soil surface due to movement of slurry N into the soil profile by infiltrating water and its loss to the atmosphere by volatilization. In general nutrient losses by surface runoff are not economically important but can be significant from an environmental standpoint.

Reduction of nitrogen losses via leaching and surface runoff

Two main ways are open to reduce N losses via leaching and surface runoff, viz. (i) by influencing the amount of mineral N in and on the soil and (ii) by influencing soil water content and flow.

Onc of the most important factors influencing the amount of mineral N in the soil is the balance between N supply and the N requirement of crops. Fertilizer applications should wherever possible be restricted to the growing season. Application outside the growing season may lead to an extra loss of nutrients by leaching and enhances the pollution risk by surface runoff.

Even when fertilizers are applied during the growing season, precipitation can cause nitrate leaching. This risk can be minimized by splitting mineral fertilizer dressings as happens on grassland and by using urea or ammonium containing fertilizers in early spring instead of nitrates. Further protection can be achieved if applications of N are restricted to levels which are sub-optimal from an agronomic standpoint. However, due to economic consequences, sub-optimal fertilization will be realized only if farmers are compensated for lost production.

Reduction of N losses by leaching and surface runoff can also be influenced by hydrological measures. Activities which lead to a lowering of the groundwater table and hence to a lower soil water content in the root zone are likely to increase nitrate leaching to groundwater. This may arise from a lower crop uptake and less removal of nitrate through denitrification in the lower part of the soil profile. Lowering of the groundwater table may result from groundwater abstraction for municipal water supply or from land drainage to improve conditions for agriculture. In contrast, the pollution risk by surface runoff is reduced by lowering the groundwater table as this increases the water storage capacity of the soil profile. In regions where the water level can be manipulated, an optimal level should be maintained by taking into account the changing risks of pollution by leaching and surface runoff. The amount of surface runoff can also be reduced by improving soil physical characteristics. Soil layers with a low permeability can be broken and mixed with other soil layers.

Most of the measures which can be taken to reduce the environmental impact of intensive grassland management have their price. Measures may result in lower dry matter yields, greater input of labour at an inconvenient time, higher energy consumption and more investment. Hydrological modification may also reduce the pollution of water with N but may conflict with interests of nature conservation and public water supply. The introduction of protective measures requires an integrated approach which takes into account all interests in a particular region.

References

1. Appelo C A J, Krajenbrink G J W, Ree C C D F van and Vasak L (1982) Modification of the groundwater quality in the recharge area of the northwest Veluwe. Bodembescherming 11, Ministerie van Volkshuisvesting, Ruimtelijke Ordening en Milieubeheer, Den Haag (in Dutch)
2. Commission of the European Community (1978) The spreading of animal excrement on utilized agricultural areas of the community; 1. Scientific basis for the limitation of quantities and criteria for rules thereon. Information on agriculture, No. 47
3. Dowdell R L and Webster C P (1974) Soil nitrogen. Fate of fertilizer nitrogen. Report Agricultural Research Council, Letcombe Laboratory, pp 55–57, London
4. Ernst L F (1973) Calculation of the transport time of flowing groundwater in the saturated zone. Nota 755, Institute for Land and Water Management Research, Wageningen (in Dutch)
5. Fonck H (1979) Nitrogen balance of some grassland fields on Cranendonck experimental farm in 1977. Nota 1137, Institute for Land and Water Management Research, Wageningen (in Dutch)
6. Fonck H (1980) Nitrogen balance of some grassland fields on Cranendonck experimental farm in 1978. Nota 1208, Institute for Land and Water Management Research, Wageningen (in Dutch)
7. Gast L F L, Taat J and Duijvenbooden W van (1985) National monitoring network groundwater quality. 2. Maps of the first sampling. Rijksinstituut voor Volksgezondheid en Milieuhygiëne, Bilthoven (in Dutch)
8. Heijden F M J van der (1977) Variation in chemical composition of the shallow groundwater below grassland. Nota 965, Institute for Land and Water Management Research, Wageningen (in Dutch)
9. Hood A E M (1976) The high nitrogen trial on grassland at Jealotts Hill. Stikstof 83–84, 395–404
10. Kolenbrander G J (1981) Leaching of nitrogen in agriculture. In Brogan J C, ed. Nitrogen losses and surface runoff from landspreading of manures, pp 199–222. The Hague: Martinus Nijhoff/Dr Junk
11. Oosterom H P (1982) Chemical composition of the upper groundwater underneath agricultural fields and forest. Nota 1385, Institute for Land and Water Management Research, Wageningen (in Dutch)
12. Oosterom H P (1984) Influence of some factors for sandy soils on nitrate leaching and transport in groundwater. An experiment with deep lysimeters. Nota 1490, Institute for Land and Water Management Research, Wageningen (in Dutch)
13. Oosterom H P and Steenvoorden J H A M (1977) Chemical composition of shallow groundwater underneath grassland of dairy farms. Nota 964, Institute for Land and Water Management Research, Wageningen (in Dutch)
14. Prins W H (1983) Limits to nitrogen fertilizer on grassland. Doctoral thesis, Agricultural University, Wageningen
15. Ryden J C, Ball P R and Garwood E A (1984) Nitrate leaching from grassland. Nature 311, 50–53
16. Rijtema P E (1980) Nitrogen emission from grassland farms – a model approach. In Prins W H and Arnold G H, eds. The role of nitrogen in intensive grassland production, pp 137–147. Wageningen: PUDOC

17. Sherwood M and Fanning A (1981) Nutrient content of surface runoff water from land treated with animal wastes. In Brogan J C, ed. Nitrogen losses and surface runoff from landspreading of manures, pp 5–18. The Hague: Martinus Nijhoff/Dr Junk
18. Steenvoorden J H A M (1981) Landspreading of animal manure and runoff. In Brogan J C, ed. Nitrogen losses and surface runoff from landspreading of manures, pp 26–33. The Hague: Martinus Nijhoff/Dr Junk
19. Steenvoorden J H A M and Fonck H (1985) Slurry on grassland. Annual Report 1984, p. 27, Institute for Land and Water Management Research, Wageningen (in Dutch)

Nitrate loss through leaching and surface runoff from grassland: effects of water supply, soil type and management

EA GARWOOD and JC RYDEN
The Animal and Grassland Research Institute, Permanent Grassland Department, North Wyke, Okehampton, Devon, EX20 2SB, and Hurley, Maidenhead, Berkshire, SL6 5LR, UK

Key words: nitrate leaching, grassland, water supply, soil type, cutting/grazing, urine-affected-areas, denitrification, slurry

Abstract. Data obtained in recent studies of nitrate leaching from grassland in the UK are reviewed. These data suggest that substantial loss of nitrogen (N) as nitrate can occur in leachate and surface runoff, particularly from swards grazed throughout the season. Under cut swards, losses are usually low (in the range 2 to 10% of the N applied). However, losses increase to as much as 50% of N applied at rates above optimum particularly when utilisation of N by the sward is restricted by soil water supply. Consistently greater losses of nitrate (25 to 40% of the N applied) occur from swards that are grazed throughout the season due largely to the return in excreta of a substantial proportion of the N ingested by the ruminant. Losses from complete grassland catchments may be lower depending on the proportion and management of cut and grazed areas. Losses from grazed grass/clover swards are markedly lower than those from grazed grass receiving substantial inputs of fertilizer N. Increased denitrification from heavy and poorly drained soils results in a lower proportion of the N being lost through leaching and surface runoff. Drainage of such soils increases the amount of nitrate lost to water courses. Leaching of nitrate following application of slurry to grassland is usually less than that occurring after application of a similar amount of N as inorganic fertilizer. This is probably due to the substantial loss of inorganic N through ammonia volatilisation immediately after application of slurry to the land surface and to the fact that as much as 50% of the N applied is in organic forms. Various strategies to reduce losses of nitrate through leaching and surface runoff are discussed.

There is increasing concern about the levels of nitrate found in supplies of drinking water in Western Europe. The high concentrations recorded in some sources are generally ascribed to leaching of nitrate from agricultural land as a consequence of the high rates of N fertilizer used on arable crops [27, 33]. Grassland has not been regarded as a major source of nitrate in water supplies except following ploughing and cultivation of land previously under grass, when substantial release of N by mineralisation can occur [40].

Grassland responds to higher rates of fertilizer N than other crops [26]. Optimum rates are in the range 350–450 kg N ha^{-1} yr^{-1} for both herbage and animal production. Rates of N application to grassland have increased considerably in the past 30 years. The average applied to 2 to 7 year leys in England and Wales has risen from 20 kg N ha^{-1} in 1951 [9] to 191 kg ha^{-1} in 1983 [8]. Much of the more intensively managed grassland receives larger applications of N. Thus in 1983, 17% received >300 kg N ha^{-1} and 5% >400 kg N ha^{-1}.

H G van der Meer, J C Ryden and G C Ennik (eds) Nitrogen Fluxes in Intensive Grassland systems.
 ISBN 978-94-010-8454-3

Table 1. Mean annual yield of N in herbage and leaching of nitrate-N from perennial ryegrass swards grown on lysimeters of a freely-drained loam between 1970 and 1982 inclusive

N applied (kg N ha^{-1})	250	500
N yield in herbage (kg N ha^{-1})[a]	234	376
N leached (kg N ha^{-1})[b]	15	145
Peak concentration of NO_3^--N in leachate, range (mg l^{-1})[b]	3–27	52–160

[a] excluding data for 1976
[b] excluding winter 1976–77

Efficiency of use of N by grass varies widely due to soil, climatic and management factors [14, 25], but decreases rapidly when the rate of application exceeds the optimum for the conditions of an individual sward particularly when conditions for uptake and utilisation by the plant are adverse. Excess N is at risk to loss by leaching and denitrification.

This paper summarises recent research on nitrate leaching from grassland in the UK. Effects of water supply and input of N on leaching losses are assessed together with the interaction between leaching and denitrification in the dispersal of nitrate from the soil profile. Particular emphasis is placed on recent work to evaluate the effect of the ruminant on nitrate leaching from grazed swards and grassland catchments.

Nitrate leaching below cut swards

Long-term lysimeter studies

Most studies of nitrate leaching from grassland have been conducted under cut swards using lysimeters. At Hurley, swards of perennial ryegrass have been maintained in 2.6 m^2 lysimeters containing a free draining sandy loam over-lying chalk since 1967. They have received either 250 or 500 kg N ha^{-1} yr^{-1} since 1970 and have been cut four times each year to simulate a continuous silage crop [15]. The amount of nitrate lost in leachate increased disproportionately with the amount of N applied (Table 1). Whereas only 6.0% of the N applied at an annual rate of 250 kg N ha^{-1} was leached each year, 29% was leached from the sward receiving 500 kg N ha^{-1}. The magnitude of these losses reflected the yields of N in herbage at different levels of N application. Annual application of 500 kg N ha^{-1} is above optimum for the soil and its location. To achieve greater efficiency of N use at this site with a mean April–September rainfall of 338 mm, supplementary irrigation normally would be required.

Data for the period 1976–1977, not included in Table 1, deserve comment. The conditions were exceptional [16]; rainfall between June

Table 2. Annual yield of N in herbage and amounts of nitrate-N leached below perennial ryegrass swards grown on monolith lysimeters of contrasting soils receiving 420 kg N ha^{-1} yr^{-1}, 1980–84 [13]

Soil series: Characteristics:	Newport Free-draining sand	Frilsham Sandy-loam overchalk	Radyr Deep loam	Tedburn Silty clay surface water gley
Available water capacity (mm)	96	137	162	177
N yield in herbage (kg N ha^{-1})				
Without irrigation	280	346	321	347
With irrigation	333	417	420	397
N loss through leaching (kg N ha^{-1})				
Without irrigation	87	43	34	58
With irrigation	64	20	14	27

and September 1975 was only 101 mm and the actual soil water deficit reached 128 mm. The yield of N in herbage was only 207 kg ha^{-1} at the lower rate of N application (250 kg ha^{-1}) and 318 kg ha^{-1} at the higher rate (500 kg ha^{-1}). Rainfall was again exceptionally low during the winter 1975–76 and only 8 and 18 kg N ha^{-1}, respectively, were leached in the 50 mm of drainage that occurred. The drought of 1976 was even more severe. There was little growth from May to August and the yield of N in herbage was only 27 and 35 kg ha^{-1}, respectively, for the 150 or 300 kg N ha^{-1} applied during this period. The substantial losses during the following winter (184 and 590 kg N ha^{-1} for the 250 and 500 kg N ha^{-1} treatments, respectively) reflect the accumulation of mineral N from fertilizer application over the two years, added to by a flush of mineralisation of soil organic N when rain fell on a still warm soil in autumn 1976.

Effect of water supply and soil texture on leaching losses

The effects of the supplementation of rainfall by irrigation and contrasting soil texture on the efficiency of utilisation of N by grass and subsequent leaching losses are shown by data from other lysimeters at Hurley [13]. These contain intact monoliths of four soils ranging from a free draining sand to a slowly permeable silty clay (Table 2). They were sown to perennial ryegrass in 1979. Since 1980, replicates have been irrigated to maintain the soil water deficit within 25 mm of field capacity; in contrast other replicates have received only natural rainfall. All received 420 kg N ha^{-1} and were cut nine times each year.

Data from these lysimeters confirm the earlier statements on the relationships between water supply, yield of N in the herbage and leaching loss of N. Annual yield of N in the herbage was least (280 kg N ha^{-1}) on the coarse textured soil of low water holding capacity which dried out rapidly in periods of low rainfall (Table 2). Although the yield of N in the herbage was increased by irrigation (333 kg ha^{-1}) this was no more than that for the heavier soils without irrigation. Losses by leaching were correspondingly greater from the free draining sand although irrigation reduced loss by 27%. Analysis of the soil solution extracted from different depths indicated little movement of N below the surface soil horizons during the growing season. This agrees with observations made by D'Aoust [10] and Sofield [32], neither of whom found significant movement of N even when excess water was applied as irrigation.

The effects of drought and irrigation on nitrate leaching from cut swards in monolith lysimeters of clay and silt loams have also been evaluated in a recent study reported by Webster and Dowdell [37]. The swards were cut at intervals of six weeks during the growing season and were fertilized with calcium nitrate at 400 kg N ha^{-1} in each of five successive years (1977–1981). Mean losses ot 41 and 15 kg N ha^{-1} were observed from the fertilized swards on the clay and silt loams, respectively, when the lysimeters were irrigated to simulate the long-term average rainfall at the sites from which the two soils were collected (670 and 730 mm for the clay and silt loams, respectively). Similar losses were observed when irrigation was used to simulate total rainfall equal to 120% of the long-term average. When drought was imposed for periods of two weeks before and after each cut or for four weeks before each cut, the total amounts of nitrate leached increased to 74 and 104, and to 33 and 109 kg N ha^{-1} for the clay and silt loams, respectively. In both drought treatments, the total annual input of rainfall plus irrigation was the same as that for the lysimeters in which the long-term average rainfall was simulated. Mean annual losses from unfertilized swards were 3.8 kg N ha^{-1}, the concentrations of nitrate in water draining below these being generally 1 mg N l^{-1}. Concentrations of nitrate increased to between 5.1 and 11.6 mg N l^{-1} when fertilizer was applied and reached 28.8 mg N l^{-1} for the four-week drought treatment.

Proportion of nitrate leached derived directly from fertilizer

Some workers [4, 12] have used ^{15}N to distinguish the extent to which leached nitrate is derived from fertilizer or soil N. In the study reported by Dowdell and Webster [12], the total amount of nitrate leached annually from lysimeter monoliths of a sandy loam sown to ryegrass receiving 400 kg N ha^{-1} yr^{-1} was in the range 8 to 33 kg N ha^{-1}.

Fertilizer contributed about 60 to 70% of the total nitrate lost in the first drainage season following application of the labelled fertilizer. The proportion of the total nitrate leached which is derived from the fertilizer applied, increases with increasing rate of application of N. Barraclough *et al.* [4] reported that 9, 39 and 75% of the nitrate leached from grassed lysimeter monoliths receiving 250, 500 and 900 kg N ha^{-1} yr^{-1}, respectively, was derived from the fertilizer applied in that year.

It is important, however, not to misjudge the significance of findings in studies using ^{15}N. Following addition of labelled fertilizer, there is rapid microbially-mediated exchange of ^{15}N with the pool of ^{14}N in the soil organic matter [20]. As the amount of ^{14}N in the pool of microbially labile organic N is large, subsequent re-mineralisation of the pool results in dilution of the ^{15}N originally immobilised. Consequently, of the total amount of N leached in a particular season only a part is derived from the fertilizer applied in that year, as illustrated by the studies described above [4, 12]. Data for ^{15}N loss must be weighed against the total loss of N and against losses from unfertilized swards. For example in the studies of Dowdell and colleagues [12, 37], in which fertilizer comprised only 48 to 70% of the total amount of nitrate leached, losses of nitrate from fertilized lysimeters were at least 10 and in some cases 100 times greater than losses from lysimeters to which no N had been applied.

Nitrate leached following fertilizer application in the spring

The data discussed above relate to annual losses of N through nitrate leaching, much of which occurs during the winter months from nitrate accumulated in the soil profile or arising from mineralisation of organic N. There is little evidence for direct loss of fertilizer nitrate during the growing season, even when irrigation is applied. However, direct leaching of fertilizer N can occur as a result of by-pass flow through soil macro-pores, particularly when heavy rainfall follows application of N in the late winter and early spring when the soil is close to field capacity. In a study reported by Barraclough *et al.* [4, 5] losses of 2.7 and 11.9 kg N ha^{-1} occurred from cut swards in early April after applications of 62.5 and 125 kg N ha^{-1}, respectively, which were followed three days later by 26.5 mm of rain. Nitrate loss from lysimeters treated with ^{15}N-labelled fertilizer confirmed that much of the nitrate leached was derived from the immediately preceding application of N. Loss of N during this single leaching event represented 58 and 22% of the annual amount of nitrate leached below these swards which received 250 and 500 kg N ha^{-1} yr^{-1}, respectively.

Impact of the ruminant on nitrate leaching

Despite the fact that the majority of grassland in temperate regions is to a greater rather than lesser extent under grazing management, most studies of nitrate leaching from grassland relate only to losses below cut swards. Such studies obscure the fate of N returned in ruminant excreta either directly during grazing or indirectly as slurry and manure from winter-housed stock. Approximately 80% of the N excreted is contained in urine and is present largely as urea. Deposition of urine during grazing may result in applications of N at rates up to and in some cases in excess of $100\,g\,N\,m^{-2}$ ($1000\,kg\,N\,ha^{-1}$) [22, 38, 39]. Although some of this N is lost as ammonia or recovered by the sward [2, 3] a substantial proportion remains in the soil profile particularly after the development of a soil water deficit. Much of this residual N is lost through a combination of leaching and denitrification. For example, Ball and Keeney [1] reported combined losses between 24 and 57% of the urine N applied at rates of 30 and $60\,g\,N\,m^{-2}$. Losses through leaching and denitrification were affected by soil and weather conditions in the period following appliction of urine; losses increased in the order warm-dry < cool-moist < warm-moist.

Ryden *et al.* [30] reported mean nitrate and ammonium contents of 53.5 and $56.3\,\mu g\,N\,g^{-1}$, respectively, in soil to a depth of 30 cm below urine-affected areas of an intensively managed ryegrass sward grazed by steers and sampled in October at the end of a season in which 420 kg $N\,ha^{-1}$ had been applied. Corresponding values for dung-affected areas were 26.2 and $13.1\,\mu g\,N\,g^{-1}$. Additional N was observed within the profile when soil below the urine-affected areas was sampled to a greater depth in November (Fig. 1). The influence of urine on the content of inorganic N in soil extended to a distance of 15 to 30 cm from the centre of the urine-affected area. At the time of sampling, nitrate was present at depths > 1 m even though there was a residual soil water deficit of approximately 15 mm at the time of sampling. Some of this nitrate may have arisen from percolation of urine through dry soil followed by hydrolysis of urea and subsequent nitrification of the ammonium produced at depth. Evidence for this is the presence of substantial amounts of ammonium at depths between 0.66 and 1.33 m in the soil profile. The mean content of inorganic N below the centre of nine similar urine-affected areas was $118\,g\,N\,m^{-2}$ ($1180\,kg\,ha^{-1}$), 77% of which was present as nitrate.

Unequivocal evidence for enhanced nitrate leaching from grazed swards was obtained in a recent study in the UK (Fig. 2) [30]. In this study, leaching was assessed from the hydrological balance at an experimental site in south-east England and from the nitrate content of soil

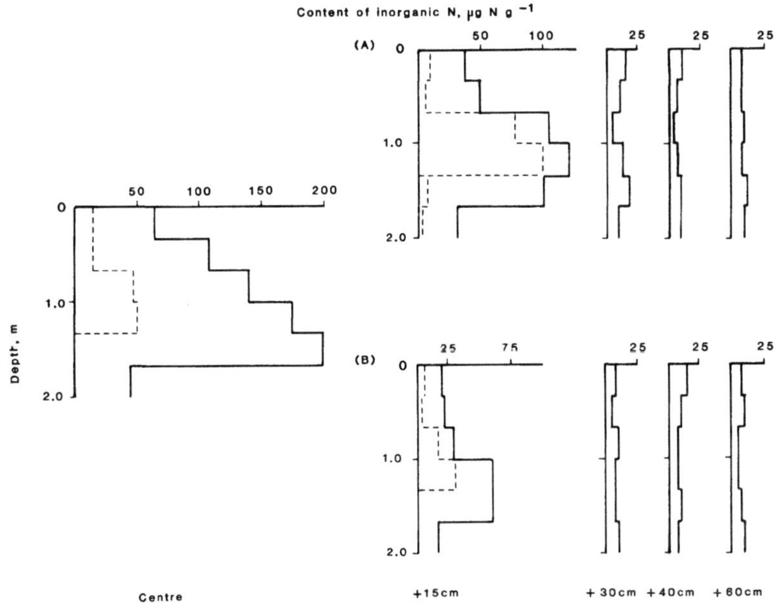

Figure 1. Distribution of nitrate (——) and ammonium (- - - -) in soil and sub-soil below the centre and at positions below axes (A and B) running at a right-angle from the centre of a urine-affected area in a grazed ryegrass sward receiving 420 kg N ha^{-1} yr^{-1}.

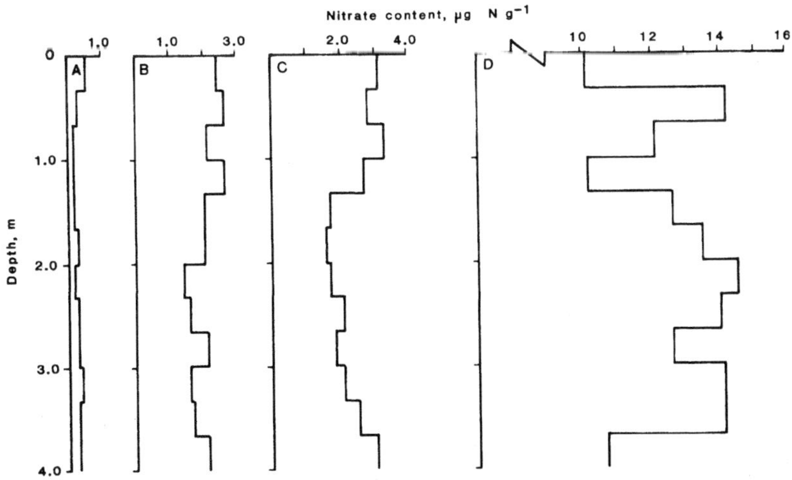

Figure 2. Nitrate content of samples of soil and chalk from successive depths below cut and grazed ryegrass/white clover swards (A and B, respectively) and below cut and grazed ryegrass swards receiving 420 kg N ha^{-1} yr^{-1} (C and D, respectively). The cut and grazed ryegrass/white clover swards fix approximately 270 and 160 kg N ha^{-1} yr^{-1}, respectively (unpublished data). All the swards were defoliated at intervals of four weeks, the grazed swards by yearling steers.

Table 3. Annual input of N to a 40 ha catchment in north-west England and losses of N as nitrate and ammonium in stream flow [19]

Year[a]	Input of N	Loss of N		Rainfall
		NO_3^--N	NH_4^+-N	
	kg N ha^{-1}	kg N ha^{-1}		mm
1972	168	40.5	0.9	1440
1973	188	40.9	5.4	1272
1974	188	35.4	6.0	1564
1975	188	27.6	3.3	1184
1976	195	34.5	5.7	1235
1977	210	55.5	9.0	1511
1978	226	70.6	8.4	1374
1979	240	73.6	7.2	1498

[a] January to December

and the under-lying chalk removed from below contrasting grazed and cut swards. The nitrate content of material removed from below each grazed sward was greater by a factor of 5.7 to 10.2 than that in material removed from below the corresponding cut sward. However, the nitrate content of material below the grazed ryegrass/clover sward was not significantly different from that below the cut ryegrass sward receiving 420 kg N ha^{-1} yr^{-1} and was 6.6 times less than that below the grazed fertilized ryegrass sward which also received 420 kg N ha^{-1} yr^{-1}. Measured rates of N fixation in the cut and grazed ryegrass/clover swards suggest annual fixation of 270 and 160 kg N ha^{-1}.

The annual losses of N through nitrate leaching were calculated from the average annual rate of percolation of water through the profile (0.8 m) and the concentration of nitrate in and the bulk density of the materials removed. These were 2.5, 23, 29 and 162 kg N ha^{-1} for the cut and grazed ryegrass/clover and cut and grazed fertilized ryegrass swards, respectively.

Loss of nitrate in leachate and surface runoff from complete grassland systems

There are few data on nitrate loss from complete grassland systems. Horne [19] reported a study conducted over a period of eight years in a high rainfall area of north-west England in which loss of nitrate arising from both leaching and surface runoff was measured in a stream draining a catchment of 40 ha. Land use within the catchment incorporated all aspects of grassland practice, namely grazing, cutting for conservation and disposal of slurry and manure. Annual loss of nitrate increased from about 35 kg N ha^{-1} to about 70 kg N ha^{-1} during a period of four years (1976–1979; Table 3) as the input of N (fertilizer plus pig slurry) increased from 188 to 240 kg N ha^{-1}. Management of the catchment was similar in each year and annual rainfall varied within fairly narrow limits. Additional loss (0.9 to 9.0 kg N ha^{-1}) occurred as ammonium-N. In contrast, loss of nitrate in a stream draining an adjacent extensively managed catchment (rough grazing for sheep) receiving no fertilizer N ranged from 5.3 to 7.9 kg N ha^{-1} [24].

Lower losses of nitrate-N through leaching and surface runoff have been observed in studies at another site in south-east England and in Northern Ireland. Hood [18] and Dowdell *et al.* [11] reported a mean annual loss of 23 kg nitrate-N ha^{-1} in drainage during six seasons from land incorporating both cut and grazed swards to which 250 kg N ha^{-1} was applied annually. This value is expected to be an under-estimate of the total loss of nitrate as the drains intercepted only 48% of the excess rainfall [11]. Hence, it is probable that much of the nitrate moving in discrete zones below urine-affected areas (e.g. Fig. 1) was not included in the measured loss. Furthermore, approximately 30% of the annual input of N was removed each year in silage whereas no slurry or manure was returned. The lower loss of N through leaching may also reflect a greater loss of nitrate through denitrification at this site as discussed below.

Jordan and Smith [21] reported a nitrate leaching loss of 17 to 36 kg N ha^{-1} yr^{-1} from a 6 ha drained catchment in Northern Ireland which received approximately 420 kg N ha^{-1} yr^{-1}, about 75% as inorganic fertilizer N, the remainder as pig slurry. The catchment comprised both cut and grazed swards, with 102 to 137 kg N ha^{-1} yr^{-1} being removed in herbage cut for conservation. The lower losses of nitrate from this intensively managed catchment may possibly reflect (i) higher rates of denitrification, (ii) the proportion of the catchment cut or grazed at different times of the year, (iii) incomplete interception of nitrate moving below urine-affected areas of the grazed swards. Again in this study, only 66% of excess rainfall was intercepted by the field drains (C. Jordan, personal communication).

A similar experiment involving hydrologically isolated continuously grazed swards has recently been initiated in south-west England [17]. The experiment comprises 12 plots, each of 1 ha, in two blocks. Within each block there are drained and undrained plots of the original sward, receiving 200 and 400 kg N ha^{-1} yr^{-1}, and of a perennial ryegrass (cv. Melle) sward sown in August 1982 and receiving 400 kg N ha^{-1} yr^{-1}. The soil at the site is a surface water gley, silty clay in texture and typical of many poorly drained grassland soils in England and Wales on which, without drainage, utilisation of grass both at the beginning and end of the grazing season is restricted. Preliminary data indicate a mean

Table 4. Nitrate-N lost in winter drainage or surface runoff from original or re-seeded grazed swards in south-west England and annual loss by denitrification from the same swards ($kg\,N\,ha^{-1}$) [17]

	N loss in surface runoff	N loss in drainage	N loss by denitrification
Original sward			
200 $kg\,N\,ha^{-1}\,yr^{-1}$:			
Undrained	16	–	n.d.[a]
Drained	1	58	n.d.
400 $kg\,N\,ha^{-1}\,yr^{-1}$:			
Undrained	55	–	64
Drained	2	160	46
Re-seeded sward[b]			
400 $kg\,N\,ha^{-1}\,yr^{-1}$:			
Undrained	36	–	79
Drained	1	79	58

[a] Not determined
[b] Sown 1 year previously

recovery of 95% for excess rainfall in drainage and surface runoff from both the drained and undrained treatments.

Substantial losses of nitrate-N in drainage were observed when high rates of N (400 $kg\,ha^{-1}\,yr^{-1}$) were applied to the drained plots of the original sward (Table 4). Losses of nitrate from the drained plots were greater than the losses in a similar volume of surface runoff water from the undrained plots of the same sward. The loss of nitrate from the drained original sward receiving 400 $kg\,N\,ha^{-1}\,yr^{-1}$ is similar to that below the grazed ryegrass sward receiving a similar application of N on a freely-drained loam in south-east England (Fig. 2). Smaller losses of nitrate were observed from the original sward receiving 200 kg N $ha^{-1}\,yr^{-1}$. Nevertheless, 29% of the N input was lost in drainage. Lower losses of nitrate in both drainage and surface runoff were also observed for the re-seeded swards. Again, drainage substantially increased the total discharge of nitrate to surface waters.

Inter-dependence of leaching and denitrification

Little attention has been given in much of the work on nitrate leaching from agricultural land to the inter-dependence of leaching and denitrification in the removal of nitrate from the soil profile. From the standpoint of water quality, denitrification may be considered the preferred pathway of nitrate loss if such loss is to occur at all. In many freely drained soils, denitrification loss at the end of the growing season is restricted initially by the development of anaerobic microsites within the profile as the soil approaches field capacity and subsequently by decreasing soil temperature [29]. Hence, the period when moderate rates of denitrification occur in such soils may extend for only a few weeks. In contrast, in less freely drained soils, denitrification may occur for extended periods at any time of the year and result in substantial removal of nitrate from the soil profile with a consequent reduction in the amount of nitrate lost through the soil profile or in surface runoff.

This inter-dependence of leaching and denitrification is illustrated by the preliminary results of the study in south-west England summarised in Table 4. Losses of nitrate through denitrification were higher in the undrained plots from which losses of nitrate to water courses were lower. Similarly, losses through denitrification in the soil below the recently re-seeded swards were higher than those in the soil below the original sward from which losses through drainage and surface runoff were considerably greater. Denitrification was measured only in the surface soil (0–10 cm) and this may account for the relatively small differences in loss between the drained and undrained treatments. The origin of differences in denitrification loss between re-seeded and original swards is as yet unidentified.

A similar inter-dependence of leaching and denitrification is illustrated in studies involving cut swards. The losses of N through nitrate leaching (6.0 and 29% of the N applied at annual rates of 250 and 500 kg N ha^{-1}, respectively) in the study described by Garwood *et al.* [14] contrast with those of 1.5 and 5.4%, respectively, of the fertilizer applied at the same rates to hydrologically isolated field plots on a less freely drained loam in the same locality [5]. Differences in the amounts of nitrate leached at each site, reflect to some extent the differences in denitrification loss from the fertilizer applied [3, 28]. Annual denitrification losses were 4 and 9% for the freely drained and less freely drained soils, respectively.

Nitrate leaching following land application of cattle slurry

Although nitrate leaching from cut grass is generally small, largely due to the efficiency of uptake of N by a growing sward, the ultimate fate of much of the N removed in herbage for conservation is determined by the utilisation of slurry that is subsequently produced. At the time of application to land, N in cattle slurry is usually distributed between inorganic and organic forms in approximately equal amounts [35], the inorganic N being present largely as ammonium. In the short term (one to two weeks after application), loss of N may occur through surface runoff at sloping sites [31] and through volatilisation of ammonia [6]. Following the short-term losses of ammonium-N, residual inorganic N will be recovered by the sward and/or nitrified. Additional nitrate will be produced as a result of the mineralisation and nitrification of a portion of the organic N originally present in the slurry. In autumn, winter and early spring when uptake of inorganic N by the plant is restricted, this nitrate is subject to loss through both leaching and denitrification.

There are few data on nitrate leaching following application of cow slurry to grassland. Burford *et al.* [7] accounted for $<1\%$ of the N applied in slurry in drainage waters collected following applications of cow slurry at rates up to 550 t ha^{-1} (approximately 2500 kg N ha^{-1}). However, it was thought that additional nitrate-N was lost in water that by-passed the field drains. In another study [34], cow slurry containing 200 kg N ha^{-1} was applied annually to ryegrass swards in spring for a period of five years. Only 10 kg N ha^{-1} was lost each year in the leachates collected at the base of lysimeters, an amount similar to that lost from a comparable sward receiving 125 kg N ha^{-1} yr^{-1} as ammonium nitrate. When four applications of slurry were made during each season, supplying 800 kg N ha^{-1}, leaching losses increased to between 15 and 30 kg N ha^{-1}. These losses were about half those from 500 kg N ha^{-1} supplied as ammonium nitrate.

The relatively small amounts of nitrate leached following even large applications of slurry N are consistent with the fact that as much as 80% of the inorganic N in slurry is lost through volatilisation of ammonia in the period immediately following surface application of slurry [6, 23]. This, combined with additional loss of nitrate through denitrification and the fact that organic N in slurry is mineralised to only a limited extent [35] results in substantially lower nitrate leaching from slurry than that observed following application of similar amounts of N as inorganic fertilizer [34]. Excessive nitrate leaching is expected only when large amounts of slurry are applied to fallow land or arable forage crops [30, 36]. However, increased interest in injection to reduce ammonia loss from slurry may result in increased leaching of nitrate, particularly from slurry injected outside the growing season. Nitrification inhibitors may have an important role in reducing the potential for such increased nitrate loss.

Conclusions

The experimental data discussed above indicate that as the management of grassland is intensified through the use of larger amounts of fertilizer N resulting in increased stocking rates, losses of nitrate through leaching and surface runoff will increase and may equal or exceed those normally occurring from arable land. Although there is wide variation in nitrate losses, under cutting management they are generally low unless the amount of N applied exceeds the optimum for the site or the amounts and timing of each application pay little regard to soil and weather conditions. Under grazing management similar criteria apply in regard to the amount of N applied and to soil and weather conditions both at the time and prior to the application of N. However, control of nitrate accumulation and subsequent loss is more complex in grazed swards as the return of animal excreta to the soil, giving localised high concentrations of N, greatly increases the potential for loss.

Current knowledge of factors affecting nitrate accumulation in the soil profile and its subsequent loss, suggests strategies to reduce these losses by increasing the utilisation and recycling of N within grassland systems. These strategies which are particularly relevant to intensive systems include:

(i) Greater integration of grazing and cutting management: (a) feeding more fresh or conserved herbage with improved storage and utilisation of slurry on cut swards; (b) taking late silage cuts from swards grazed earlier in the season to utilise accumulated nitrate-N. If conditions are dry, irrigation would increase plant uptake of N residual from the grazing period.

(ii) More carefully managed application of N during the growing season to reduce the accumulation of nitrate within the soil profile. Continued application of N in dry conditions before previously applied N has been utilised or late into the autumn will lead to large losses, particularly from swards grazed throughout the season.

(iii) Use of strategic irrigation to enable efficient uptake of N applied in mid to late season.

(iv) Increased use of white clover in grazed swards with strategic applications of fertilizer N particularly in late winter/early spring.

References

1. Ball P R and Keeney D R (1983) Nitrogen losses from urine-affected areas of a New Zealand pasture, under contrasting seasonal conditions. In Smith J A and Hays V W, eds. Proceedings of the XIV International Grassland Congress, (Lexington), pp 342–344. Boulder, Colorado: Westview
2. Ball P R, Keeney D R, Theobald P W and Nes P (1979) Nitrogen balance in urine-affected areas of a New Zealand pasture. Agronomy Journal 71, 309–314
3. Ball P R and Ryden J C (1984) Nitrogen relationships in intensively managed temperate grassland. Plant and Soil 76, 23–33
4. Barraclough D, Geens E L and Maggs J M (1984) Fate of fertilizer nitrogen applied to grassland, II. Nitrogen-15 leaching results. Journal of Soil Science 35, 191–199
5. Barraclough D, Hyden M J and Davies G P (1983) Fate of fertilizer nitrogen applied to grassland, I. Field leaching results. Journal of Soil Science 34, 483–497
6. Beauchamp E G, Kidd G E and Thurtell G (1982) Ammonia volatilization from liquid dairy cattle manure in the field. Canadian Journal of Soil Science 62, 11–19

7. Burford J R, Greenland D J and Pain B F (1976) Effects of heavy dressings of slurry and inorganic fertilizers applied to grassland on the composition of drainage waters and the soil atmosphere. In Agriculture and Water Quality, Technical Bulletin No. 32, pp 432–443, Ministry of Agriculture, Fisheries and Food. London: H.M.S.O.
8. Church B M (1984) Use of fertilizers in England and Wales, 1983. Rothamsted Report for 1983, pp 295–300. Harpenden, UK: Rothamsted Experimental Station
9. Church B M and Lewis D A (1977) Fertilizer use on farm crops in England and Wales: Information from the survey of fertilizer practice 1942–1976. Outlook on Agriculture 9, 186–193
10. D'Aoust M J (1965) Some aspects of the interaction between nitrogen and water in the growth of grass swards. Ph.D. thesis, University of Reading
11. Dowdell R J, Morrison J and Hood A E M (1980) The fate of fertilizer nitrogen applied to grassland: uptake by plants, immobilisation into soil organic matter and losses by leaching and denitrification. In Prins W H and Arnold G H, eds. The role of nitrogen in intensive grassland production, pp 129–136. Wageningen: PUDOC
12. Dowdell R J and Webster C P (1980) A lysimeter study using nitrogen-15 on the uptake of fertilizer nitrogen by perennial ryegrass swards and losses by leaching. Journal of Soil Science 31, 65–76
13. Garwood E A and Roberts D (1985) The potential for grass production of differing soil types and nutrient losses from these soils with and without irrigation. The Grassland Research Institute, Annual Report 1984. Hurley, UK: The Animal and Grassland Research Institute
14. Garwood E A, Salette J and Lemaire G (1980) The influence of water supply to grass on the response to fertilizer nitrogen and nitrogen recovery. In Prins W H and Arnold G H, eds. The role of nitrogen in intensive grassland production, pp 59–65. Wageningen: PUDOC
15. Garwood E A and Tyson K C (1973) Losses of nitrogen and other plant nutrients to drainage from soil under grass. Journal of Agricultural Science, Cambridge, 80, 303–312
16. Garwood E A and Tyson K C (1977) High loss of nitrogen in drainage from soil under grass following a prolonged period of low rainfall. Journal of Agricultural Science, Cambridge, 89, 767–768
17. Garwood E A, Tyson K C, Denehy H L, Stone A C, Reid T C, Ryden J C, Armstrong A, Atkinson J L and Hallard M (1985) The effects of field drainage on sward productivity and utilisation, soil physical condition and nutrient movements. The Grassland Research Institute, Annual Report 1984. Hurley, UK: The Animal and Grassland Research Institute
18. Hood A E M (1976) Nitrogen, grassland and water quality in the United Kingdom. Outlook on Agriculture 8, 320–327
19. Horne B (1980) Soil, water and fertilizers. Great House Experimental Husbandry Farm, Annual Review 1980, pp 21–26. London: Ministry of Agriculture, Fisheries and Food
20. Jansson S L and Persson J (1982) Mineralisation and immobilisation of soil nitrogen. Nitrogen in agricultural soils, pp 229–252. Madison, Wisconsin: Soil Science Society of America
21. Jordan C and Smith R V (1985) Factors affecting leaching of nutrients from an intensively managed grassland catchment in County Antrim, N. Ireland. Journal of Environmental Management (in press)
22. Kolenbrander G J (1981) Leaching of nitrogen in agriculture. In Brogan J C, ed. Nitrogen losses and surface runoff from landspreading of manures, pp 199–221. The Hague: Martinus Nijhoff/Dr Junk
23. Lauer D A, Bouldin D R and Klausner S D (1976) Ammonia volatilization from dairy manure spread on the soil surface. Journal of Environmental Quality 5, 134–141
24. Ministry of Agriculture, Fisheries and Food (1981) ADAS Research and Development Reports. Grassland, fodder conservation and forage crops, pp 1–2
25. Morrison J (1980) The influence of climate and soil on the yield of grass and its response to fertilizer nitrogen. In Prins W H and Arnold G H, eds. The role of nitrogen in intensive grassland production, pp 51–57. Wageningen: PUDOC
26. Morrison J, Jackson M V and Sparrow P E (1980) The response of perennial ryegrass to

fertilizer nitrogen in relation to climate and soil. Report of the Joint GRI/ADAS Grassland Manuring Trial – GM 20. Technical Report 27. Hurley, UK: The Animal and Grassland Research Institute
27. The Royal Society (1983) The nitrogen cycle of the United Kingdom, a Study Group Report. London: The Royal Society
28. Ryden J C (1983) Denitrification loss from a grassland soil in the field receiving different rates of nitrogen as ammonium nitrate. Journal of Soil Science 34, 355–365
29. Ryden J C (1986) Gaseous losses of nitrogen from grassland (this volume)
30. Ryden J C, Ball P R and Garwood E A (1984) Nitrate leaching from grassland. Nature 311, 50–53
31. Sherwood M and Fanning A (1981) Nutrient content of surface runoff water from land treated with animal wastes. In Brogan J C, ed. Nitrogen losses and surface runoff from landspreading of manures, pp 5–17. The Hague: Martinus Nijhoff/Dr Junk
32. Sofield I (1980) A computer simulation model of the soil- plant- and N-relationships in cut grass swards. Ph.D. thesis, University of Reading
33. Tomlinson T E (1971) Nutrient losses from agricultural land. Outlook on Agriculture 6, 272–278
34. Unwin R J and Smith K A (1983) Fertiliser value of organic manures in the UK. Proceedings No 221. London: The Fertiliser Society
35. Van Dijk T A and Sturm H (1983) Fertilizer value of animal manures on the Continent. Proceedings No 220. London: The Fertiliser Society
36. Vetter H and Steffens G (1981) Leaching of nitrogen after the spreading of slurry. In Brogan J C, ed. Nitrogen losses and surface runoff from landspreading of manures, pp 251–271. The Hague: Martinus Nijhoff/Dr Junk
37. Webster C P and Dowdell R J (1984) Effect of drought and irrigation on the fate of nitrogen applied to cut permanent grass swards in lysimeters: leaching losses. Journal of the Science of Food and Agriculture 35, 1105–1111
38. Whitehead D C (1970) The role of nitrogen in grassland productivity. Bulletin 48, 202 p. Farnham Royal, UK: Commonwealth Bureau of Pastures and Field Crops
39. Whitehead D C (1986) Sources and transformations of organic nitrogen in intensively managed grassland soils (this volume)
40. Young C P and Gray E M (1978) Nitrate in groundwater – the distribution of nitrate in chalk and Triassic sandstone aquifers. Technical Report TR69, 66 p. Marlow, UK: The Water Research Centre

GPSR Compliance

The European Union's (EU) General Product Safety Regulation (GPSR) is a set of rules that requires consumer products to be safe and our obligations to ensure this.

If you have any concerns about our products, you can contact us on ProductSafety@springernature.com

In case Publisher is established outside the EU, the EU authorized representative is:

Springer Nature Customer Service Center GmbH
Europaplatz 3
69115 Heidelberg, Germany

Batch number: 10607224

Printed by Printforce, the Netherlands